Samuel Hubbard Scudder

Brief Guide to the Commoner Butterflies of the Northern United States and Canada

Being an introduction to a knowledge of their life-histories

Samuel Hubbard Scudder

Brief Guide to the Commoner Butterflies of the Northern United States and Canada
Being an introduction to a knowledge of their life-histories

ISBN/EAN: 9783337248468

Printed in Europe, USA, Canada, Australia, Japan

Cover: Foto ©Andreas Hilbeck / pixelio.de

More available books at **www.hansebooks.com**

BRIEF GUIDE TO THE COMMONER BUTTERFLIES OF THE NORTHERN UNITED STATES AND CANADA

Being an Introduction to a Knowledge of their
Life-Histories

BY

SAMUEL HUBBARD SCUDDER

NEW YORK
HENRY HOLT AND COMPANY
1893

ROBERT DRUMMOND, ELECTROTYPER AND PRINTER, NEW YORK

PREFACE.

During the preparation of a long-projected and still unpublished Manual of the Butterflies of North America, it occurred to me that when that was ready there would still be needed something less technical; something which should introduce to the young student the names and somewhat of the relationships and lives of our *commoner* butterflies; and that if such a guide were restricted to the commoner butterflies of the region where it would be most used, viz., our Northern States east of the Great Plains—much the same territory as was originally and wisely covered by Gray's Manual of Botany—the actual extent of the work would be so limited as to bring it within the reach of all, not alarm the beginner by its magnitude, and, because they are better known, permit a fuller account of their interesting life-histories.

I have accordingly selected the butterflies—less than a hundred of them—which would almost surely be met with by any industrious collector in the course of a year's or two years' work in the more populous Northern States and in Canada, and have here treated them as if they were the only ones found there. I have omitted many species which are common enough in certain restricted localities (such, for instance, as our White Mountain butterfly) and included only those which are common over wide areas. As the earlier stages of these insects are just as varied, as interesting, and as important as the perfect stage, descriptions are given of these under the guidance of the same principle, only such stages as would be more commonly met with being fully described, and the egg and earliest forms of caterpillar omitted as rarities and as also too difficult for the beginner's study. If, then, a young student can find nothing in this work to correspond with his particular capture, then he may rest assured that it is not one of the more common kinds, and he will have to go to the larger and more technical works to discover what it is. At any rate, he is likely to be pleased: either he has found out what it is and can thereby learn something of what is already known about it; or he has found a rarity, a discovery not always distressing to the amateur.

To aid in these determinations, separate keys are appended for each of the three stages, caterpillar, chrysalis, and butterfly, by which any insect included in the work may be tracked.

There is another advantage in this restriction of the work to the commoner butterflies, for these are better known in the various stages of their lives, and interest in them is thereby greatly enhanced. I should be loath indeed to treat of butterflies as if they were so many mere postage-stamps to be classified and arranged in a cabinet; and if, by adding to the mere descriptions of the different species in their various most obvious stages some of the curious facts concerning their periodicity, their habits of life, and their relations to the world around them, I may spread before the eyes of the young some of the attractions which lie at the open door of Nature and induce some to wander into the by-ways for more eager personal search, I shall have gained my end.

Those wishing still further accounts of the different species here described, and particularly descriptions and figures of the egg and earlier stages of the caterpillar of any one of them, are referred to my "Butterflies of the Eastern United States and Canada," and to Edwards's "Butterflies of North America," in one or the other of which ample accounts will often be found.

Species which are found in the region embraced in this work, but not regarded as sufficiently common therein to merit a place in it, are mentioned by name in their appropriate places in smaller type; they number just about as many as those of which descriptions are given, and full accounts of most of them will also be found in the works above mentioned.

A short Introduction to the study of Butterflies in general, with special application to our own, is prefixed to the body of the work, and is followed by a brief section showing where the principal literature upon the subject is to be found. An explanation of some of the terms used is appended, and a figure added on p. 60 explanatory of the nomenclature of the wing.

CAMBRIDGE, April 13, 1893.

CONTENTS.

	PAGE
Preface	iii
Introduction	1
What are Butterflies?	1
The Structure of the Perfect Insect or Imago	2
The Appearance of the Egg	5
What the Caterpillar is like	6
The Character of the Chrysalis	7
A Few Words about the Eggs	8
The Lives and Habits of Caterpillars	9
How the Chrysalis Hangs	12
The General History of Butterflies	14
Variation in the Butterfly	15
Some Remarkable Differences between the Sexes	20
The Senses of Butterflies	22
Mimicry and Protective Resemblance	23
The Classification of Butterflies	25
Some Works on American Butterflies	27
Keys to the various Groups	33
Key to the Groups, based on the Perfect Butterfly	34
Key to the Groups, based on the Caterpillar	45
Key to the Groups, based on the Chrysalis	53
Nomenclature of the Parts of the Wing	60
The Commoner Butterflies of the Northern United States and Canada	63
Family Brush-footed Butterflies	63
Subfamily Danaids	63
Genus Anosia	63
Anosia plexippus	63
Subfamily Nymphs	66
Tribe Crescent-Spots	66
Genus Euphydryas	66

CONTENTS.

	PAGE
Euphydryas phaeton	66
Genus Cinclidia	68
Cinclidia harrisii	68
Genus Charidryas	69
Charidryas nycteis	69
Genus Phyciodes	71
Phyciodes tharos	71
Tribe Fritillaries	72
Genus Brenthis	72
Brenthis bellona	72
Brenthis myrina	74
Genus Argynnis	76
Argynnis atlantis	76
Argynnis aphrodite	77
Argynnis alcestis	78
Argynnis cybele	79
Genus Speyeria	80
Speyeria idalia	80
Genus Euptoieta	81
Euptoieta claudia	81
Tribe Angle-Wings	82
Genus Junonia	82
Junonia coenia	82
Genus Vanessa	84
Vanessa cardui	84
Vanessa huntera	85
Vanessa atalanta	87
Genus Aglais	89
Aglais milberti	89
Genus Euvanessa	90
Euvanessa antiopa	90
Genus Eugonia	92
Eugonia j-album	92
Genus Polygonia	93
Polygonia progne	93
Polygonia faunus	94
Polygonia comma	95
Polygonia interrogationis	97
Tribe Sovereigns	98
Genus Basilarchia	98
Basilarchia arthemis	98

CONTENTS.

	PAGE
Jasoniades glaucus	148
Genus Euphoeades	150
Euphoeades troilus	150
Genus Heraclides	151
Heraclides cresphontes	151
Genus Papilio	153
Papilio polyxenes	153
Family Skippers	155
Tribe Larger Skippers	155
Genus Epargyreus	155
Epargyreus tityrus	155
Genus Thorybes	156
Thorybes pylades	156
Genus Thanaos	158
Thanaos lucilius	158
Thanaos persius	159
Thanaos juvenalis	161
Thanaos brizo	162
Thanaos icelus	163
Genus Pholisora	164
Pholisora catullus	164
Genus Hesperia	165
Hesperia montivaga	165
Tribe Smaller Skippers	166
Genus Ancyloxipha	166
Ancyloxipha numitor	166
Genus Atrytone	167
Atrytone zabulon	167
Genus Erynnis	169
Erynnis sassacus	169
Genus Anthomaster	170
Anthomaster leonardus	170
Genus Polites	170
Polites peckius	170
Genus Thymelicus	171
Thymelicus mystic	171
Genus Limochores	173
Limochores taumas	173
Explanation of some Terms	175
Appendix: Instructions for Collecting, etc.	179

CONTENTS.

	PAGE
Tribe Coppers	127
Genus Chrysophanus	127
Chrysophanus thoe	127
Genus Epidemia	128
Epidemia epixanthe	128
Genus Heodes	128
Heodes hypophlaeas	128
Genus Feniseca	130
Feniseca tarquinius	130
Family Typical Butterflies	132
Subfamily Pierids	132
Tribe Red-Horns	132
Genus Callidryas	132
Callidryas eubule	132
Genus Zerene	133
Zerene caesonia	133
Genus Eurymus	134
Eurymus philodice	134
Eurymus eurytheme	135
Genus Xanthidia	137
Xanthidia nicippe	137
Genus Eurema	138
Eurema lisa	138
Genus Nathalis	139
Nathalis iole	139
Tribe Orange-Tips	140
Genus Anthocharis	140
Anthocharis genutia	140
Tribe Whites	141
Genus Pontia	141
Pontia protodice	141
Genus Pieris	143
Pieris oleracea	143
Pieris rapae	144
Subfamily Swallow-Tails	145
Genus Laertias	145
Laertias philenor	145
Genus Iphiclides	146
Iphiclides ajax	146
Genus Jasoniades	148

CONTENTS.

	PAGE
Basilarchia astyanax	101
Basilarchia archippus	102
Tribe Emperors	104
Genus Anaea	104
Anaea andria	104
Genus Chlorippe	105
Chlorippe clyton	105
Chlorippe celtis	106
Subfamily Meadow Browns or Satyrs	107
Genus Cissia	107
Cissia eurytus	107
Genus Satyrodes	108
Satyrodes eurydice	108
Genus Enodia	109
Enodia portlandia	109
Genus Cercyonis	110
Cercyonis alope	110
Cercyonis nephele	111
Family Gossamer-winged Butterflies	113
Tribe Hair-Streaks	113
Genus Strymon	113
Strymon titus	113
Genus Incisalia	114
Incisalia niphon	114
Incisalia irus	115
Incisalia augustus	116
Genus Uranotes	117
Uranotes melinus	117
Genus Mitura	118
Mitura damon	118
Genus Thecla	119
Thecla liparops	119
Thecla calanus	120
Thecla edwardsii	121
Thecla acadica	122
Tribe Blues	123
Genus Everes	123
Everes comyntas	123
Genus Cyaniris	125
Cyaniris pseudargiolus	125

INTRODUCTION.

1. What are Butterflies?

One of the great groups or "orders" into which insects are divided is called Lepidoptera (derived from two Greek words meaning scaly-wings). This group differs from all other insects by having in the perfect stage a long, hollow, thread-like tongue, through which fluids may be sucked or rather pumped up, and which, when not in use, is coiled up like a watch-spring; and by having four rather broad wings covered with colored scales overlying one another in rows like shingles, slates, or tiles on a roof. These insects undergo striking changes in the course of their lives; for they are hatched from the egg as crawling worms having a globular head with biting jaws, and a body supported not only by the three pairs of short horny legs found in the young of most insects, but by several, generally five, pairs of stumpy, fleshy legs behind them; while the two joints of the body next following those with horny legs and some other joints near the hinder end never have any; from this they change into a pupa or chrysalis, a mummy-like object with the legs, wings, and other members swathed upon the breast and with no possible motion except in the wriggling of the joints of the abdomen or hinder end of the body; from this temporary prison escapes in due time the winged creature of beauty which adds such a charm to the summer landscape.

Butterflies differ from other Lepidoptera by having clubbed or knobbed antennæ in their perfect stage, and generally in their transformations, for most of them are hung up by silken cords attached to hooks on the tail, and sometimes also by a girth around the waist; they are rarely enclosed in cocoons, or, if so, the chrysalis is in most cases also supported within; while moths (i.e., all other Lepidoptera) usually construct silken cocoons, often of very close texture, or make cells in the ground, in either of which cases the chrysalis lies loosely within or attached by the tail only. Butterflies usually fly by day, moths usually by night. Butterflies usually rest with their wings erect; moths usually with wings flatly expanded or sloping downward on either side like a tent.

2. The Structure of the Perfect Insect or Imago.

The body of a butterfly is distinctly separated into three divisions: the head, to which the antennæ and the coiled tongue are attached; the chest, trunk, or thorax, which supports the four wings and three pairs of legs; and the abdomen.

The head is the smallest part, but contains a wonderful lot of interesting organs. The sides are almost entirely occupied by large faceted eyes; from the summit spring a pair of slender thread-like but apically clubbed antennæ; while beneath, between the scaly and hairy upcurved three-jointed appendages, called labial palpi, the spiral tongue (maxillæ) is coiled.

The most interesting of these organs is this tongue. It coils up just like a watch-spring, but may be extended at full length, as when plunged into the depths of a flower in search of honey. It appears as if single and solid, but is really composed of two exactly similar lateral halves grooved along their inner surface, so that when placed together the opposing grooves form a fine tube; and to secure them in place, so that the tube shall not leak, the edges of the grooves are delicately notched so as to dovetail into corresponding teeth on the edge of the opposing groove, by which they become closely interlocked.

To enable the butterfly to pump into its body through this tube the honeyed sweets of flowers, the throat at the base of the tube expands into a sac with muscles radiating toward the walls of the head and others encircling it; when the first set of muscles contracts, the interior space of the sac is enlarged; when the encircling muscles contract, it is diminished. By the alternating action of these sets, a pumping process goes on aided by a little flap at the base of the tube which lets the fluids pass in but not out; so that the squeezing of the full sac presses the fluids into the stomach; its enlargement creates a vacuum which causes the honey in the flower to ascend the tube past the valve into the sac.

The antennæ may be divided into a base consisting of two joints stouter than those beyond; a thread-like stalk, slender and equal, consisting of many joints; and the club, which is composed of the swollen tip, sometimes arising almost insensibly from the stalk, sometimes abruptly; and in the Skippers having usually a recurved hook at the tip; the club is usually at least twice as thick as the middle of the stalk, generally naked beneath and often flattened.

The eyes are usually very convex, but vary in different groups in this respect as well as in the amount of space they cover; they are ordinarily naked, but sometimes delicately hairy, and in the Skippers are overhung by a curving tuft of bristles. The number of facets in the eye is very great, numbering thousands to each eye.

The thorax is divisible into three parts, called from in front backward prothorax or fore-trunk, mesothorax or mid-trunk, and metathorax or after-trunk. The prothorax, however, is scarcely more than a flattened plate in front, and is easily overlooked; the division between the other two masses is readily seen behind when the scales are rubbed off, and the mesothorax is seen to be much the largest part of the thorax.

The fore wings are attached to the mesothorax, the hind pair to the metathorax, and both are composed of two films supported by a system of branching hollow rods and the surface covered with scales.

Of these rods there are ordinarily four or five to each wing, but when all are present there are six. The two middle ones of the six are the only ones that branch, and are called respectively the subcostal (the upper one) and the median; generally they meet or nearly meet near the middle of the wing and enclose what is called the discoidal cell, and the subordinate rods or nervules appear to diverge from its margin.

The scales are hollow flattened sacs, covered with longitudinal striæ on the upper surface and generally toothed or serrate at the tip, with a short bulbed stem by which they are fixed in the wing membrane; upon which they lie like shingles on a roof, and by their pigment and the refraction of light by their surface striæ give to the wing all its color and delicate markings.

Certain scales, however, are peculiar to the male sex and are curiously distributed in special patches or concealed positions so as scarcely to be visible even under the microscope until they have been uncovered. These are often fringed with tassels at the end, each thread of the tassel a canal leading through the body of the scale to a gland at the base and so serving as scent-organs—the odors being sometimes appreciable to human senses and then in all known cases agreeable perfumes like flowers, sandal-wood, and musk.

The legs are six in number, one pair to each division of the thorax; they are always very slender and stick-like. The front pair, however, as we pass from the lower to the higher butterflies becomes more and more atrophied and useless, first in the males, then in the females, until in the highest family they are utterly useless, often not easy to detect, and render this group practically four-legged instead of six-legged.

Their principal divisions are the femur (plural, femora) or thigh, the tibia or shank—these two parts generally of about equal length and indivisible; and the tarsus, the last composed of five always unequal joints, armed beneath with short spines and at tip with claws, a pad, and often with paronychia or whitlows, a sort of membranous imitative accompaniment of the claws, perhaps best seen in the Pierids.

The abdomen is formed of nine essentially simple segments. The males may be distinguished from the females by the structure of the last segment, the females being provided with a pair of minute flaps, one on each side, which protect and form part of the ovipositor, while the males have side clasps and an upper median hook for clasping the body of the female. The abdomen of the female when filled with eggs is very much larger and fuller than that of the male, and the sex can thus often be told at a glance.

3. THE APPEARANCE OF THE EGG.

The eggs of butterflies are very various in sculpture, and though often very simple, are at other times exquisitely ornamented. They are usually broad and flat at the base, and more or less rounded above. One class may be called, in general, barrel-shaped; but this would include minor divisions, such as thimble-, sugar-loaf-, flask-, or acorn-shaped, or even fusiform; others are globular, or hemispherical, or tiarate. The surface may be more or less deeply pitted, or delicately reticulate, or broken up by vertical ribs connected by raised cross lines, or may be perfectly smooth and uniform; but all have a collection of microscopic cells at the centre of the summit perforated by little pores, forming the micropyle, through which the egg is fertilized; and these microscopic parts are often of exceeding beauty.

4. What the Caterpillar is like.

Caterpillars of butterflies do not differ from those of moths by any single characteristic. Each family of Lepidoptera has certain peculiarities, and one has to become more or less familiar with them to determine whether or not a given kind falls in this or that family.

They are worm-like creatures, but with a distinct horny head, separable from the body.

The head is very different from that of the future butterfly, having biting jaws, no compound eyes, but in their place a semicirclet of simple ocelli, and antennæ hardly visible without a glass; these last, indeed, are very like the palpi, a series of two to four rapidly-diminishing rounded joints ending in a bristle.

The body is composed of thirteen (apparently twelve) segments of which the first three, corresponding to the joints of the future thorax, have each a pair of horny five-jointed legs ending with a single claw; while the third to sixth and last abdominal segments bear each a pair of two-jointed fleshy "prolegs," armed at tip with a single or double series of minute hooklets. Breathing pores or spiracles are found on the sides of the first thoracic and the first eight abdominal segments. Besides this, the whole body is clothed, when adult, with short hairs or longer spines set on little pimples, or with fleshy filaments or tubercles of some sort, all arranged to a greater or less extent (excepting generally the short hairs) in longitudinal series, but these are often not precisely aligned on the thoracic and abdominal segments.

In their earliest stage, however, before their first moult and sometimes for a stage or two after it, the clothing of the caterpillar is very different from what it is at maturity, the appendages usually consisting at first of longer or shorter bristles, often tubular and conveying fluids to the enlarged summit, and arranged in longitudinal series different from those of the spines or filaments of the mature caterpillar. This earliest stage, therefore, needs special attention in the study of butterflies, although the creature is then exceedingly minute, and, therefore, not considered in the present work.

Certain caterpillars (and this peculiarity usually runs through whole groups of allied forms), have certain glands opening externally which may secrete fluids or odors of various kinds; some of these are eversible like the Y-shaped appendages on the top of the segment behind the head of the Swallow-Tails and here termed "osmateria"; or the lateral polypiform extrusions called "caruncles" on both sides of one of the hinder segments of some of the Blues, both kinds of organs being thrown out only under provocation,

5. The Character of the Chrysalis.

In this state the creature is a sort of mummy, all the appendages, both of head and thorax, folded over upon the breast, packed closely and tightly glued, extending usually to the fourth abdominal segment. In a few of the lower butterflies, the tongue extends still further and is then more or less free. All of the appendages, however, are not seen, for the palpi and hind legs are entirely concealed beneath the other members, and the organs that appear are ranged in the following order from the middle line outward: tongue, fore legs, middle legs, antennæ, fore wings, hind wings, of the latter of which very little is seen, they being mostly covered by the fore pair.

The body is compact, but there are usually some marked prominences upon the surface, notably in certain places, such as the front of the head, which usually has a pair of projections, sometimes only one; the middle of the back of the mesothorax, often ridged or with a pointed projection; the extreme base of each of the wings, which are usually tuberculate or humped; and the middle line of the back of the abdomen or the sides of the same, which are often ridged. In the highest family, where the caterpillars are spined, there are often rows of conical tubercles on the chrysalis corresponding generally to the position of the larger spines of the caterpillar.

This is all that need be said regarding the actual structure of butterflies in their different stages to one beginning their study, for it is better to dwell rather upon their lives and protean changes, their histories and habits, if we wish to gain a true and favorable insight into their characteristics.

6. A Few Words about the Eggs.

The eggs of butterflies are always laid in full view, excepting that in a few instances they are partially concealed by being thrust into crevices. Ordinarily they are laid on one or the other surface of the leaves of the food-plant of the caterpillar or on the stem of the same, and usually on or in contiguity to the tenderer growing leaves. As a general rule, the eggs are laid singly, in some instances on the extreme tip of a pointed leaf; but in not a few cases they are laid in clusters of from two or three to several hundreds. Sometimes these are rude bunches piled loosely or in layers one upon another; sometimes they are laid in more or less regular single or double rows; sometimes in a single column of three or four or even as many as ten eggs, one atop another; or they may girdle a twig like a fairy ring. The duration of the egg state is commonly from one to two weeks, but it varies in different species in the summer-time from five or even less days to about a month; there are, however, some butterflies which pass the winter in the egg state. In all such cases the eggs are laid upon the stem, never upon the leaf, and some spot is chosen, like the neighborhood of a leaf-scar, which affords a certain amount of protection during the winter.

7. THE LIVES AND HABITS OF CATERPILLARS.

When eggs of butterflies are laid in clusters, the caterpillars are almost invariably social to a greater or less degree, at least in early life, sometimes to maturity ; if they are laid singly and it is only by accident that several are laid near together, the caterpillars are solitary. In the majority of cases where the egg is laid singly, the first act of the escaping caterpillar is to devour it entirely or in greater part.

Solitary caterpillars may live exposed on the upper or the under sides of leaves, or they may retire to the stem of the food-plant for greater security, or they may construct, each for itself, some kind of concealment, or live within fruits. When fully exposed, they usually remain quite motionless, stretched at full length when not feeding, and may select for their resting-place peculiar spots. The most curious is one adopted by some Brush-footed Butterflies (and the egg is then commonly laid at or near the extreme tip of the leaf) which devour the apical portion of the leaf, leaving the midrib untouched, and perch themselves upon this midrib after having attached to it by a few threads a small packet of bits of leaf and frass which is moved by every breath of wind,—probably to distract the attention of its enemies from itself.

Others construct shelters more or less complicated. Some merely spin transverse threads across the floor of a leaf, causing its sides to curl, and then recline, half hidden, in the shallow trough; others make it so complete that the edges meet and the leaf forms a cylinder; still others fasten the opposite edges by silk and by biting weaken the resistant ribs and also the main rib so that the leaf droops; others bite channels into the leaf at two distant points and turn the flap thus formed over upon the leaf, securing it in place by silken strands; while for winter use the partly grown caterpillar of the later brood of Basilarchia and some allied genera not only coils a leaf into a cylinder but lines it within and without with silk, leaves a ledge to crawl out upon, and secures the leaf to the twig by strong silken fastenings. In nearly all these cases the caterpillar seems to rest upon the upper surface of a leaf and curl the sides upward, very rarely the reverse.

But there are others which fasten several leaves together, generally very slightly, to form a leafy bower, or in the case of grasses a tubular burrow; and in a few instances, as in *Vanessa huntera*, bits of the inflorescence of the plant are caught in the slight meshes of the net to make a more perfect concealment. Among our Larger Skippers many which live half their life in a nest formed of a single leaf finish it in a bower made of many.

Social caterpillars often construct nests in company, which then often embrace in an irregular web the whole or nearly the whole of a branch of the food-plant. Usually the web is thin and hardly conceals the surface, but sometimes it is almost like parchment, as in the Mexican *Eucheira socialis*. Winter is sometimes passed in one of these webs, and when constructed, as it sometimes is, on an annual, the shrinkage after the death of the stalk makes a compact mass of leaves, frass, web, and caterpillars, from which it would seem as if no caterpillar could escape in the spring. When social caterpillars construct no shelter, they usually feed side by side in rows, and move from place to place in files.

A very large number of our caterpillars live through the winter, and this is often the only means by which a species survives the inclement season; most of them hibernate when about half grown; others, strange to say, just from the egg, without having eaten anything but the shell from which they came; still others hibernate full grown and full fed, changing to chrysalis just when vegetation starts in the spring. Some of these caterpillars, especially those partly or fully grown, construct nests for hibernation; others use the same nest which has served their larval life, strengthening it against the greater needs of winter; others seek crannies of any kind.

In some cases where the caterpillars of a second brood hibernate when half-grown, the caterpillars of the first brood at the hibernating age, but in midsummer, will fall into lethargy, from which some will arouse after say a fortnight's quiescence, while others will prolong their premature into actual hibernation, and in the following spring caterpillars of the same stage but of two successive broods will mingle together.

It is apparent, then, that there is considerable variety in the duration of life of caterpillars. Instances are on record where the time from birth to chrysalis was only about ten days; ordinarily it is at least a month; with those that hibernate it may be in some cases nearly a year;

while there are several instances known where caterpillars have lived over two winters and might therefore take from eighteen to twenty or more months for their larval existence alone.

8. How the Chrysalis Hangs.

In making its preparation for its final moult, when the change to chrysalis is to take place, the caterpillar proceeds in exactly the same manner as in preceding moults, except that it spins more silk and, in addition to the carpet on which it stands, adds other strands of a special nature, according to the method in which the chrysalis is to swing. The chrysalis is provided with special hooks at its posterior end with which to engage the silken pad prepared for it, excepting in the case of a few which change on the surface of the ground.

One mode of suspension is to hang pendent by the tail alone from a pad of silk. Generally free to swing with every jar or breeze, the more so as the pad is usually more or less loosely woven, there are some in which the hooks are distributed over a more or less elongated area, and, the caterpillars having constructed a more compact pad, the attachments are firmer and more extended, so that the chrysalis may be more or less rigid and even hang in a position by no means vertical but inclined strongly toward the horizontal.

The movements of chrysalids of the pendent type are not confined to the looseness of attachment of the hooks or the nature of the web to which they cling, but in all there is more or less capability of motion by the sliding of the abdominal joints upon one another, and the chrysalis may thus effect voluntary motion, sometimes, when disturbed, of an extraordinarily active kind. Some chrysalids, moreover, make slow periodic diurnal movements, heliotropic or phaotropic, i.e. toward or away from the sun or light, sometimes lateral, sometimes forward and backward.

Other chrysalids are attached not only by the tail but also by a girth, whether tight or loose, slung around the middle of the body in the dorsal depression or saddle which always exists between the thoracic and abdominal regions. If the girth be tight, the ventral surface of the chrysalis, which touches the surface of rest, is nearly or quite straight; if loose, it is often bent to a greater or a less degree opposite the girth, or describes a curve with the same point as the middle of the arc.

A modification of this mode of suspension is seen in some Skippers, which make cocoons in which both the median girth and sometimes to a less extent the tail attachments form Y-shaped strands, which are attached at their extremities to the walls of the cocoon; into the centre of one set the hooks of the tail are plunged, while the middle of the body is slung between the longer arms of the other and larger set of strands.

There is but one family of butterflies in which all the members construct cocoons—the Skippers. Their cocoons are usually of a rather fragile nature and consist (usually) of leaves, blades of grass, or other vegetable material, generally living, shaped into a more or less oval or cylindrical cell by silken attachments; sometimes the interior is more or less perfectly lined with a thin membrane of silk; within this, as just stated, the chrysalis hangs by means of Y-shaped shrouds, the form of the smaller one sometimes difficult to determine from the mingling of its threads with those forming the extremity of the cocoon.

Chrysalids which give birth to butterflies the same season vary in their duration from about three days to a month, but usually from ten days to a fortnight. But a considerable number pass the winter in this shape, and may then endure from five to eleven months, and sometimes this latter variation may occur in a single species having several broods, in which an increasing proportion of each successive brood of chrysalids of one season pass over the ensuing winter. Instances are on record in which chrysalids, normally hibernating, have been known to pass over a second winter and then give birth to the butterfly.

9. THE GENERAL HISTORY OF BUTTERFLIES.

Beginning life as an egg which usually hatches within a few days after being laid, the young caterpillar finds its sole duty to be to eat and escape being eaten. It feeds voraciously, and outgrows its skin so often that it is obliged to moult four or five times before it is full grown. On each of these occasions it stops feeding for a while, spins a carpet of silk, and fastens its claws therein; when the time for change comes, the old skin splits along the middle of the back of the thoracic segments by violent muscular effort, the old head-case (from which the new head was first withdrawn) is shaken off and the creature crawls out of its old skin, which in many instances it thereupon devours. In the last change, to chrysalis, the head is not removed from the old skin, but itself splits in the middle and down one or both sides of the frontal triangle, and the chrysalis emerges. After hanging awhile, the chrysalis skin splits at much the same points and the butterfly emerges to begin the cycle again with the laying of eggs.

The cycle of changes through which a butterfly moves is in temperate climates commonly passed once each year,— or rather once each season, for it is winter that usually interferes with the activities by robbing the creature of its means of sustenance and paralyzing its action. Inasmuch as the pupal stage is in the higher insects the period of longest inactivity, one would presume beforehand that this period would coincide with winter; and so it does in a large number of cases. Yet among butterflies the exceptions to such a rule are not only exceedingly common, but, as might be expected were there any departure, they are very varied and winter is passed, by one species or another, in every conceivable stage of existence, including every part of caterpillar life. Indeed, cases are not unknown, especially in high latitudes and altitudes, where more than one season is required to bring a butterfly to maturity.

On the other hand, a large number of our butterflies, and this is especially true southward, complete the cycle of their changes twice or oftener in a season, and there are not a few having an extended latitudinal range which vary in this respect, having one or more broods in the northern part of their range, and an added brood or more in the southern. The end of the season generally surprising multiple-brooded butterflies in all stages of existence, an opportunity has easily arisen for every possible form of hibernation or lethargic life, which accounts for the variation discoverable in the lives of our butterflies, each form settling at last upon that series of changes which is best fitted for it.

10. VARIATION IN THE BUTTERFLY.

Like most creatures, butterflies, when they are found over a wide territory, show great difference between individuals found in the extremes of the range, so that it is sometimes difficult to tell, at least until collections are made over the intervening country, whether specimens from distant places should be regarded as distinct species or as geographical varieties. The most skilled may make mistakes for lack of proper material.

But quite apart from this, butterflies appear to be exceptionally sensitive to the environment and to offer an unusual amount of variation of a different sort; for dimorphism or polymorphism of various kinds, that is, the existence of a given species under recognizably distinct forms (two or more, even sometimes to five or six) is by no means uncommon.

This distinction is often sexual; indeed there are relatively few species in which the outward aspect of the two sexes does not differ, in some cases to a remarkable degree. It is universal in the numerous species of Eurymus, for example, where in general the inner margin of the dark outer bordering of the wings is sharp and precise in the male, confused and irregular in the female. In very many cases, however, it is accompanied by a simple dimorphism, sometimes affecting one sex only (and then usually the female), as in many species of Eurymus, where one form of female has the bright ground color of the male, the other a pallid ground color; at other times affecting both sexes, as in some species of Polygonia : in *P. interrogationis*, for example, there are four sets of individuals differing in the general coloring of both surfaces of the wings and even in the form of the wings—differences all of which may occur in the progeny of a single individual and fed on the same plant.

But these differences are very often correlated with, generally confined to, differences of brood. One of the most striking and at the same time one of the simplest examples is in the double-brooded European species *Araschnia prorsa*, where the first brood is composed of individuals of one type with highly variegated markings (levana), the second of a very distinct type with more sharply-contrasted coloring (prorsa), which, until they were bred from each other, were universally, and reasonably, regarded as distinct species. This is called seasonal dimorphism.

Numerous striking examples occur in this country, not a few of which are excellently shown in Edwards's Butterflies of North America, such as many species of Polygonia (in *P. interrogationis* they are largely seasonal, the latest brood being all of one type), *Phyciodes tharos*, the species of Pieris, and especially *Iphiclides ajax*. The latter instance is the more remarkable, because the three forms (marcellus, telamonides, and ajax), though sequent in the order named, do not strictly represent distinct broods, since the earlier emerging individuals of the first brood are marcellus, the later-appearing individuals of the *same* brood are telamonides, while the subsequent broods, of which there are several, are ajax.

Distinct climatal differences, whether temperature or moisture (or both), are unquestionably the prime cause of seasonal dimorphism, the former in temperate, the latter in tropical, regions. The first has been practically proved by experiment, the latter by the correspondence of the phenomena to that of temperate climates and their synchronism with the dry and wet seasons.

Many cases of dimorphism are compound. Instances of this have already been given; indeed, most cases of dimorphism involve some distinct element, such as season or latitude, or temperature in some form. Thus, *Jasoniades glaucus*, which exhibits dimorphism in the female, does so only in the south, for the dark form of the female (in which the conspicuous normal stripes of the male are obscured) occurs but rarely north of Pennsylvania, although there is a distinct tendency in both sexes to a broadening of the darker markings and the partial suppression of the yellow in high northern latitudes or their equivalent, as among the White Mountains of New Hampshire. A similar instance occurs in *Everes comyntas* with the boundary limits of the dark female at about the same place.

Nearly all the above instances of dimorphism where it is not of the simplest kind (whether seasonal or not) may be termed polymorphic, since more than two types of individuals appear in a single species; especially is this the case where a sort of double dimorphism occurs, like that of *Iphiclides ajax* or of *Polygonia interrogationis* mentioned above. Instances have also been cited where the geographical element entered; but polymorphism is most conspicuous and complicated where all the above elements are combined,—where dimorphism between the sexes, dimorphism also between the members of one sex confined to distinct portions of the range of the species, and seasonal dimorphism more or less limited in its geographical range and in its correlation with the broods (as the species may be multiple-brooded or not), may be further complicated by geographical variations independent of and running through all the others. Two cases may be cited as remarkable instances of complicated polymorphism if the facts shall prove well grounded.

In the extreme north, *Cyaniris pseudargiolus* is single-brooded and appears in two forms, an earlier with heavier markings (lucia) and a later (violacea); the males of both are blue above; the females paler blue with broad dark margins to the fore wings. In New England it is double-brooded, the sexes differing as before; the first brood is trimorphic and serial, the earliest individuals having heavy markings (lucia), the next intermediate markings (violacea), the last light markings (neglecta), while the second brood is composed entirely of neglecta; in the northern part of the belt in which the first brood is trimorphic, the form neglecta is comparatively rare, and lucia the most abundant, while the reverse is the case in the southern part of the same belt (and lucia itself is so variable that one type of it has been separated as marginata). Farther south lucia disappears altogether and the first brood is dimorphic,—violacea and neglecta in the order of their appearance; but now a new element is introduced, for the males of violacea become dimorphic, one form resembling the males of the same found farther north, the other being uniformly dark above (violacea-nigra). In the southern part of its range, the latest individuals (neglecta) of the first brood are usually much larger than the members of the second brood, all of which are otherwise of the same type. This butterfly flies not only from Hudson Bay to Georgia, but also from the Atlantic to the Pacific, and in California we have a new form (piasus), hardly distinguishable from neglecta, which appears to be double-brooded in the south but to show no difference between the broods. Farther north, however, near the British boundary, the conditions of New England are at least in part repeated, while in Arizona an ashen variety (cinerea) occurs.

The different forms assumed by *Eurymus eurytheme* have caused their description as distinct species on four or five occasions. It, too, has an immense range. In Texas the cycle begins in November (the summer and not the winter interfering with its activities) with a yellow type (ariadne) succeeded by a yellow-orange type (keewaydin) and finally by an orange type (amphidusa), each a distinct brood, the last-named indeed double-brooded; with the increase of temperature, the size and the depth and brilliancy of color increase; the form keewaydin has a sexually dimorphic female, one resembling the male in ground color, the other pallid (keewaydin-pallida), and the form amphidusa is similarly favored (amphidusa-alba). In the northern part of the range of the species, the earliest (May) form, a yellow one, differs so much from the earliest (November) type of the south as to be given a distinct name (eriphyle), and when keewaydin and amphidusa have had their turn, it again appears in the latter part of the season, and though the autumn form has not received a distinct name, it can be distinguished from the spring form, at least in the male sex, the spring individuals being uniform chrome yellow above, while the October males are of a whitish yellow and the hind wings are dusted with gray.

11. Some Remarkable Differences between the Sexes.

Many male butterflies may be readily distinguished by characteristic tufts, rows, or wisps of hairs or patches of special scales or membranous folds generally rendered in some way conspicuous, and which do not occur in the female. Of the first we have a good example in our species of Argynnis, which show a row of long semi-recumbent hairs on the upper surface of the hind wings between the costal and subcostal nervures; of the second in the mealy-looking margins of the upper surface of the wings of Callidryas, the discal patch on the fore wings of many Hair-streaks, the apparently blackened and thickened veins of the fore wings of Argynnis, or the discal streak accompanied by large tilted scales so common in the Smaller Skippers; of the last in the blackened pocket of the hind wings of Anosia, the plaited fold of the hind wings of Laertias, or the deftly inconspicuous costal fold of the Larger Skippers.

These very patches or folds usually conceal scales differing to a greater or less extent from the surrounding scales and peculiar to the males, called scent-scales or androconia, i.e., male-scales. They do not, however, always occur in these patches (where they are usually concealed from view to some degree), but may be simply scattered among the other scales and then, being almost invariably much smaller, almost completely concealed from view.

While the ordinary scales of butterflies, common to both sexes, show very little variety in their structure, being striate, more or less fan-shaped or shingle-shaped laminæ with finely-toothed apical margin, the androconia show an extraordinary variety of structure, but are rarely toothed at the tip. They may be shaped like an Indian club, a shepherd's crook, a long needle ending with a whip-lash, a twisted ribbon, a battledore, an elongated fan, a row of beads, a spatula, a tapering ribbon with fringed tip, or may assume many other forms which could only be described at length; they are generally very slender and minute. Where they are fringed, it is highly probable that the separate threads of the fringe are so many canals conducting to glands at the base of the scale, for in many instances odors plainly perceptible have been traced to this source.

These odors are in all cases of an agreeable nature and have generally been compared either to the fragrance of certain flowers or to the musky odors of quadrupeds; the last is a very common scent among insects and is known in such different creatures as the imago of the beetles Prionus and Osmoderma, the imago of the butterfly Argynnis, and the half-grown caterpillar of the moth Arctia parthenos.

These androconia are very capricious in their occurrence both as to exact location and as to their presence or absence in allied forms. They appear to be almost invariably present in all the species of any given genus or else absent from all, but allied genera in a single tribe often vary in this particular. They occur in all families and in most, perhaps all, tribes of butterflies.

They are usually found upon the upper surface of the fore wings, very rarely, if ever, upon the under surface of any; they may be scattered indiscriminately over the wing, be collected into definite but vague areas traversing the interspaces, assemble along the principal nervures or at the extremity of the discal cell, or in a narrow discal streak or costal fold, or be confined to a little pocket on the broad face of the hind wings, or lie in a closed plait next the anal margin, or in various other positions.

12. The Senses of Butterflies.

The power and range of vision in butterflies (and in insects in general) have without doubt been popularly overestimated. Both direct experiments and study of the structure of the compound eye lead to the same conclusion: that while insects have a quick perception of moving objects or of objects among which they are moving, they have no power of distinguishing precise form or delicate distinctions of color or patterns, their visual perception being confused or vague.

The delicacy of the sense of smell in insects, and especially in Lepidoptera, makes full amend for defective vision. The quick advent of males among many tribes to secluded and concealed females, the possession of many odoriferous organs, the evidence that many others exist where the odors are imperceptible to human sense, all point to a delicate and keen perceptive power in this direction. It is altogether probable—and no other explanation has so great probability—that it is by the exercise of this sense that the parent butterfly discovers the proper food-plant for the deposition of her eggs. The organs for this sense are probably resident in the antennæ.

The fondness of butterflies for the honeyed sweets of flowers at once suggests a high development of the sense of taste; for that it is not purely a matter of hunger or the need of nourishment may be seen in the cases so often noted where butterflies fill their bodies until they can scarcely fly, which is far beyond any need of nourishment; or in the groups which continue for hours around a moist spot in a road imbibing the innutritive fluids. The organs for this sense are probably resident in the tongue-papillæ.

There seem to be no reasons for believing that any high degree of power in hearing is to be found among butterflies, as there are no organs known to serve as receptive elements, and the sounds made by butterflies are apparently due simply to the rustling of the wings. All motions that look as if possibly meant to convey sound (where none can be detected by the human ear), such as the quivering of the wings in sexual approximation, may be solely to waft emitted odors the more effectively.

Little can be said or presumed regarding touch of animals whose external parts are all crustaceous; but it is plain that warmth and cold, which deal with the same nervous elements, have decided influences in every stage beyond the egg. The ordinary inactivity of caterpillars in the night can not be laid to the absence of light, for their behavior in darkened apartments is much the same as out of doors; the movements of chrysalids tell the same story; and we know that a measurable amount of movement of the antennæ occurs with changing temperature in hibernating, practically dormant, butterflies.

13. MIMICRY AND PROTECTIVE RESEMBLANCE.

Most butterflies when at complete rest close their hind wings back to back and sink the fore wings as far as possible into concealment behind them. The area of these wings then exposed to view is in a very large proportion of butterflies so colored and mottled or marbled as to render the butterfly immensely less conspicuous in its resting-place than if settled with wings expanded or the front pair not mostly concealed; in very many cases so little conspicuous as to be difficult to detect. Rarely are any other parts similarly colored. That this resemblance is protective there can be no doubt, especially in view of its common occurrence.

There are, however, innumerable instances of special and striking provisions in this direction, of which one of the most generally known is that of the oriental genus Kallima, the species of which are highly colored on the upper surface and conspicuous objects when in flight, but which are so colored and marked upon the under side that when alighted upon a twig, as they do with the fore wings thrown well forward and all wings closed, the pattern and color of the under surface are such as to make a perfect resemblance to a leaf whose midrib, a colored stripe crossing both wings and terminating at the apex of the fore pair, takes its rise from a tail-like extension of the hind wings which just reaches the twig from which the mock leaf thus springs, the tail of the wing corresponding to the pedicel of the leaf !

These phenomena, however, reach their culmination in the examples of mimicry of one butterfly by another, of which there are numerous examples of an extraordinary kind such as perhaps no other group of animals can produce. A large proportion of the objects of mimicry belong to the subfamily Euplœinæ, known to be a group protected to a large extent against foes by the possession of nauseous qualities, and it is therefore presumed that all other objects of mimicry have from some cause or other some immunity from early death above their fellows. Such a supposition is the only one, and a sufficient one, to account for the extraordinary resemblance of otherwise unprotected butterflies, especially in the female sex (for not always do the males become mimickers), to such nauseous or protected butterflies, a resemblance not only striking for its exceedingly impressive quality, but for the departure required from the normal type of coloring or pattern of the group, or even from that of the other sex alone; for its extension to structural features, such as length of antennæ and form of wing, and to mode of flight; and also for the fact that the mimicker seems to fly only in the territory occupied by the mimicked, while in neighboring territory occupied by another of the protected group another mimicker more nearly resembling it will represent it. We have one remarkable example of this mimicry in our own country in the resemblance of *Basilarchia archippus* to *Anosia plexippus*.

14. THE CLASSIFICATION OF BUTTERFLIES.

The number of family groups into which butterflies should be primarily divided has been variously given by naturalists as from two to sixteen. Writers who have insisted on any large number have, however, relied mainly upon single and relatively unimportant characters, mostly drawn from the neuration of the wings of the imago, and almost ignoring the earlier stages of the insects. Those who have paid serious attention to the latter and have regarded all parts of the structure have generally considered the number as from four to six. In the present work they are regarded as but four in number, called Skippers (Hesperidæ), Typical Butterflies (Papilionidæ), Gossamer-winged Butterflies (Lycænidæ), and Brush-footed Butterflies (Nymphalidæ).

If we examine these different groups with regard to their interrelationship it is plain that the Skippers show by far the greatest and most numerous points of resemblance to the moths; and if we look to the sum of the characters of each as regards their departure from the characteristics of the lower Lepidoptera, we shall see that they unquestionably fall into the order in which they are here placed. In addition to this we shall find two very distinct parallel series in structure and transformations which follow precisely the same course, each independent of the other, each pointing out the lines along which development has proceeded and thus indicating a natural classification.

One of these concerns the mode of transformation. In the moths, with very few exceptions, a cocoon or cell is formed within which the transformations take place. The Skippers form a cocoon, but lighter than is common among the moths, and in addition (perhaps not universally, but very generally) the chrysalids are loosely swung up within the cocoon by the Y-shaped shrouds mentioned above. The Typical Butterflies retain the shrouds though they drop the cocoon, but, as the result, the hinder shrouds become a mere pad of silk, the median shrouds a loose loop. The only change in the Gossamer-winged Butterflies is the tightening of the median loop and the flattening of the ventral surface of the chrysalis to correspond. Lastly in the Brush-footed Butterflies the median loop is dropped and the chrysalis hangs by the tail-fastenings alone, while the straight ventral surface is generally retained—a significant atavistic indication of the girt stage.

The other regards the structure of the forelegs of the imago. In the Skippers these agree perfectly with the other legs (as in the moths), except in the presence of a median spine on the tibiæ. The same is true of the Typical Butterflies excepting that the median spine is wanting in one of the two subfamilies (Pierids) regarded as the further removed from the Skippers. In the Gossamer-winged Butterflies atrophy has begun, but is insignificant excepting in the male sex. While in the Brush-footed Butterflies atrophy in both sexes has extended to complete disuse in both, though usually more excessive in the male than the female; one subfamily, nearest to the Gossamer-winged Butterflies, partakes in this particular of the characters of the latter, namely, the Snout Butterflies or Long Beaks (Libytheinæ).

SOME WORKS ON AMERICAN BUTTERFLIES.

The first important work on American Butterflies was published in England nearly a century ago by Sir Edward Smith, and contained the observations and colored illustrations of John Abbot, an Englishman some time resident in Georgia. The work * was issued in two folio volumes, but only a part of the first volume treated of butterflies, the remainder relating to moths. Drawings of caterpillar, chrysalis, and butterfly were given in every case, and as a rule they were very well executed. A single page of text accompanied each plate, and 24 plates of butterflies are given, representing as many species. Many unpublished drawings of Abbot are still preserved, as he supported himself by their sale and was a most industrious entomological artist.

The first substantial addition to our knowledge, so far as the early stages are concerned, was derived principally from the same source — Abbot's drawings. This was a smaller octavo volume † prepared by Dr. Boisduval of Paris in collaboration with Major LeConte of New York, published in parts but never completed. The twenty-six parts contained 78 plates, illustrating about 93 species, while the text only covered 85 species, not all of which

---

* The Natural History of the rarer Lepidopterous Insects of Georgia. 2 vols. fol. London, 1797. 104 pl.

† Histoire générale et iconographie des Lépidoptères et des chenilles de l'Amérique septentrionale. 8vo. Paris, 1829–42. 228 pp., 78 pl.

were figured. The illustrations, in color, are inferior to those of the preceding work. Both the above works can now be obtained only by chance through the second-hand dealers of Europe, and are high-priced.

Two other richly illustrated and costly works upon our native butterflies have been published in our own country. The first is Edwards's Butterflies,* a serial work, irregularly issued and of which the third volume is now nearly completed. The plan of this work is to describe and figure rare or interesting species or those of which the life-history has been discovered, the species following no regular order. Usually only a single species is given on a plate, but sometimes two or more of one genus appear, or a species may cover two or three plates. The wealth, delicacy, and accuracy of the drawings in certain species has never been surpassed or even nearly equalled in any work ever published in any country; nowhere else have the eggs, caterpillars, and chrysalids of single species or the variations of the perfect butterfly been illustrated with such copiousness; while the text is often full of the most interesting accounts of the habits and life of the insects. Each volume contains 50 plates or more, and on the 162 which have appeared up to this writing about as many different butterflies have been depicted; of 57 of the species more or less abundant details of the early stages are given and often a surprising number of illustrations. Through this work the early lives of some of our butterflies are better known than those of any other country, and this often applies to species from far-distant and inaccessible parts of the country like the Rocky Mountains. Nearly all the illustrations are in color.

The other work is of a more limited scope, but has the advantage of completeness as far as it goes, and of a systematic arrangement whereby our knowledge becomes

---

* The Butterflies of North America. 3 vols. 4to. Boston, 1868-93.

clearer.* It covers only the butterflies of nearly the same region as the present volume, but describes them all, and as far as possible in every stage of life with exceptionally full accounts of their distribution and life-histories, and full definitions of the characters of all the higher groups as well as of the species, drawn from every stage of life. 164 species are described, and some account of the early stages is given of all but 35 of them, of many far fuller details than ever before.

Two other books published a generation or more ago on the insects of limited regions may be mentioned, because they gave particular attention to our butterflies. The first † was by Emmons, describing such species as he knew from New York and giving figures of them. This work contained a bare description of the perfect butterflies (31 species), and colored illustrations (occupying the part or whole of 6 plates) poorly engraved and colored; it contained nothing new and was very poorly executed. It is not now of the least value.

Quite otherwise is the less pretentious but classic work of Harris,‡ which, though purporting to treat only of injurious insects and mainly those of Massachusetts, contained in the last edition (to a far less extent in the earlier editions of 1841 and 1852) descriptions and figures of a number of New England butterflies as defoliators of trees, etc., including descriptions of some new forms ; 54 species are described, and, when known,—which was not then the case with many,—brief descriptions are given of the earlier

---

* The Butterflies of the Eastern United States and Canada. By S. H. Scudder. 3 vols. imp. 8vo. Boston, 1889. 44+1958 pp., 96 pl., of which 41 are colored.

† The Agriculture of New York, Vol. V. 4to. Albany, 1854. 8 + 272 pp., 50 pl.

‡ A Treatise on some of the Insects injurious to Vegetation. 3d ed. 8vo. Boston, 1862. 640 pp., 278 figures, 8 col. pl.

stages and full accounts of the habits, perhaps half of the text being given up to these latter features. The figures, 54 of them, are, with 9 exceptions, woodcuts and remarkable examples of the woodcutter's art, all being engraved by Henry Marsh.

Two other books of my own may be mentioned here, since they deal largely with the life-histories of our butterflies. The first * is based upon a course of lectures upon butterflies in general, and has something in particular to say about 74 of our butterflies, with figures illustrative of many of them. The other † treats in the fullest possible manner of the structure, life-history, distribution, and habits of a single butterfly, *Anosia plexippus*, at every point drawing comparisons with others, so that it serves in a measure as a popular introduction to all.

Finally, attention may be directed to three or four works which deal almost exclusively with the butterfly stage and give descriptions either of all our known species or of all found in a definite portion of our country. The first ‡ pretends to be nothing but a compilation of published descriptions (many of them translations from the French) arranged in a systematic order, preceded by a very meagre key to the genera. It contains 240 species, but is now quite out of date.

The second § is an original systematic description of the

---

* Butterflies : their Structure, Changes and Life-histories, with special reference to American forms. 12mo. New York, 1881. 10 + 322 pp., 201 figs.

† The Life of a Butterfly. 16mo. New York, 1893. 186 pp., 4 plates.

‡ Synopsis of the described Lepidoptera of North America, Part I. Diurnal and Crepuscular Lepidoptera. Compiled by J. G. Morris. 8vo. Washington, 1862. 27 + 368 pp.

§ The Butterflies of the Eastern United States, for the use of classes in zoology and private students. By G. H. French. 12mo. Philadelphia, 1886. 402 pp., 93 figs.

butterflies of the same region as the present work, but including also the Southern States east of the Mississippi. 201 species are included in the work, which is preceded by an analytical key for the determination of the species, but which is largely based on color; the genera are nowhere characterized except in this key, and there too vaguely or scantily to be of much assistance. The early stages are treated of only under the species, the descriptions being compiled and condensed from preceding writers.

The third * is called a manual and covers the whole North American field north of Mexico; but it is difficult to understand how it can well be used as such, as it consists of bare descriptions of the species, with scarcely the slightest aid to discovering the genera; consequently one may have to wade through the whole to find the one sought. Its redeeming features are the cuts, which, though very rude, are generally confined to some characteristic part, a single wing or even a part of a wing. 625 species are given, and each of the woodcuts contains several figures. The plates are exceedingly poor. No attention whatever is paid to the early stages. The work reflects no credit upon the author beyond his industry. Nor does an earlier work, by the same,† on New England butterflies, in which an attempt is made to characterize the genera and higher groups and some little attention is given to the caterpillars and chrysalids; for the work is so filled with errors as to be quite untrustworthy, and the figures so very poor as to be available only when the butterfly has little resemblance to any other; when most needed they are of least use.

The histories of our butterflies, however, are by no

---

* A Manual of North American Butterflies. By C. J. Maynard. 8vo. Boston, 1891. 4 + 226 pp., 60 figs., 10 pl.
† The Butterflies of New England. 4to. Boston, 1886. 4 + 68 pp., 8 pl. col.

means related only in the works we have mentioned. Others are spread broadcast in all manner of places and only the diligent student can find them. The greater portion of these scattered accounts will be found in the miscellaneous writings of Henry Edwards, W. H. Edwards, Fitch, Fletcher, Gosse, Lintner, Riley, Saunders, and Scudder; and are particularly to be looked for in the pages of the different entomological publications of our country, past and present, and especially in the "Canadian Entomologist," "Psyche," and "Papilio."

KEYS TO THE VARIOUS GROUPS.

IN using the following keys the student has only to keep in mind three points:

1. That there are always two contrasting alternates to choose from (occasionally three).

2. That these alternates are marked by similar initial letters, A, B, c, d, etc., and by similar indentation on the page, and distinguished by superior numerals, A^1, B^2, c^3, etc.

3. That the contrasting alternate is the nearest line in the same set which begins with the same indentation and the same initial letter, though with a different numeral.

For example, in the first table, the A^1 on p. 34 has its alternate A^2, which is a long way off (on p. 42), but is nevertheless the next line beginning with an A, and it has the same indentation, while D^1 on p. 34 is immediately followed by D^2.

When alternates relate, one or the other or both of them, to tribes or higher groups, an initial *capital* is prefixed; when both refer to genera, or pairs of genera, a *small letter* is prefixed. The final terms are the numbered genera.

For the explanation of the numbered veins in the first table, see the figure on p. 60.

Key to the Groups, Based on the Perfect Butterfly.

A^1. Antennæ near together at base, less than half as far apart as the height of the eye, the end clubbed but not hooked; eyes with no overarching pencil of bristles.
B^1. Resting on four legs only, the fore legs being unused, much shorter than the others, without claws at the end, and folded against the breast.
(Fam. Brush-footed Butterflies.)
C^1. None of the veins of fore wings swollen at the base.
D^1. Antennæ without any scales.
(Subfamily Danaids.) 1. *Anosia*.
D^2. Antennæ covered, at least above, with numerous scales. ......(Subfamily Nymphs.)
E^1. Club of antennæ short and stout, three or more times as broad as the stem, more or less abruptly thickened.
F^1. Naked portion of club of antennæ with only a single longitudinal ridge or none.
G^1. Club of antennæ about three or four times as long as broad; palpi slender, compact, the last joint from one third to one half as long as middle joint.
(Tribe Crescent-Spots.)
h^1. Middle joint of palpi of nearly equal size throughout; fore tibia of male stout and swollen, not more than five or six times longer than broad.

i^1. Outer margin of fore wing scarcely shorter than the hind margin.
　　　　　　　　2. *Euphydryas*.
i^2. Outer margin of fore wing much shorter than the hind margin.
　　　　　　　　3. *Cinclidia*.
h^2. Middle joint of palpi tapering considerably on apical half; fore tibia of male very slender and of equal size throughout, at least ten times longer than broad.
i^1. Last joint of palpi nearly half as long as the middle joint; fore tibia of male much shorter than the femur.
　　　　　　　　4. *Charidryas*.
i^2. Last joint of palpi less than one third as long as the middle joint; fore tibia of male scarcely shorter than the femur.......... 5. *Phyciodes*.
G^2. Club of antennæ spoon-shaped, about twice as long as broad; palpi large and bushy, the last joint extremely short ....(Tribe Fritillaries.)
h^1. Vein 2^3 of fore wings arising before the end of the cell.
i^1. Middle joint of palpi more than three fourths longer than the greatest length of the eye......7. *Argynnis*.
i^2. Middle joint of palpi only about one fourth longer than the greatest length of the eye.......8. *Speyeria*.
h^2. Vein 2^3 of fore wings arising beyond the end of the cell.
i^1. Curve of outer margin of fore wings opening outwardly......6. *Brenthis*.

i^2. Curve of outer margin of fore wings
opening inwardly ....9. *Euptoieta*.
F^2. Naked portion of club of antennæ with three
distinct longitudinal ridges.
(Tribe Angle-Wings.)
g^1. Fore wings rounded in the interspace
between 2_1 and 2_2.
h^1. Eyes naked ; conspicuous eye-like spots
on fore wings above...10. *Junonia*.
h^2. Eyes hairy ; no conspicuous eye-like
spots on upper surface of fore wings.
11. *Vanessa*.
g^2. Fore wings sharply angulated in the interspace between 2_1 and 2_2.
h^1. Basal three fifths of hind wings uniformly dark above; no silvery comma
in middle of hind wings beneath.
i^1. Hinds wings without spinous hairs on
under surface..........12. *Aglais*.
i^2. Hind wings with numerous straight
spinous hairs beneath.
13. *Euvanessa*.
h^2. Basal three fifths of hind wings above
more or less spotted with black;
centre of hind wings beneath with a
white or silvery comma-like mark.
i^1. Hind border of fore wings straight.
14. *Eugonia*.
i^2. Hind border of fore wings strongly
sinuous ....15. *Polygonia*.
E^2. Club of antennæ long and slender, hardly
more than twice as broad as the stem,
gradually thickened.
F^1. Club of antennæ with four longitudinal
ridges on naked portion ; vein 0 of

hind wings arising opposite the parting of veins 1 and 2.
(Tribe Sovereigns.) 16. *Basilarchia*.
F^2. Club of antennæ with three longitudinal ridges on naked portion; vein 0 of hind wings arising beyond the parting of veins 1 and 2.
(Tribe Emperors.)
g^1. Antennæ fully as long as the width of the fore wings .... 18. *Chlorippe*.
g^2. Antennæ much shorter than the width of the fore wings .... 17. *Anæa*.
C^2. Some of the veins of the fore wings swollen at the base.
(Subfam. Satyrs or Meadow-Browns.)
d^1. Antennæ gradually thickened from just beyond the middle .... 19. *Cissia*.
d^2. Antennæ gradually thickened only on the apical third or fourth.
e^1. Eyes hairy.
f^1. Tibial spines of middle legs very numerous; antennæ composed of less than 36 joints .... 20. *Satyrodes*.
f^2. Tibial spines of middle legs infrequent; antennæ composed of more than 40 joints................ 21. *Enodia*.
e^2. Eyes naked .... 22. *Cercyonis*.
B^2. Resting on six legs, the fore legs, however, sometimes a little shorter and with diminished armature, at least in the male.
C^1. Of small size. Face between eyes much narrower than high; eyes notched to give room for the antennæ.
(Fam. Gossamer-winged Butterflies.)
D^1. Vein 2^3 of fore wings simple; under side of hind

wings generally with continuous markings....(Tribe Hair-Streaks.)
 e^1. Hind wings without thread-like tails.
  f^1. Hind wings of very different shape in the two sexes, the outer border not crenulate................23. *Strymon.*
  f^2. Hind wings of similar form in the two sexes, the outer border crenulate.
   24. *Incisalia.*
 e^2. Hind wings with one or two thread-like tails.
  f^1. Interspace of hind wings between veins 4 and 5 apically lobed; male with no stigma on fore wing.........25. *Uranotes.*
  f^2. Interspace of hind wings between veins 4 and 5 not produced; male with stigma on fore wing above.
   g^1. Club of antennæ comparatively short and stout, only five times as long as broad................26. *Mitura.*
   g^2. Club of antennæ comparatively long and slender, eight times as long as broad.
    27. *Thecla.*
D^2. Vein 2^3 of fore wings forked; under side of hind wings generally with discontinuous markings.
 E^1. Spines on under side of tarsi comparatively few and ranged in pretty regular series; colors of upper surface usually more or less violet and dark brown.
  (Tribe Blues.)
  f^1. Hind wings with thread-like tails. 28. *Everes.*
  f^2. Hind wings without tails.....29. *Cyaniris.*
 E^2. Spines on under side of tarsi numerous and clustered irregularly at the sides; colors of upper surface more or less

KEY TO THE GROUPS—BUTTERFLY. 39

coppery or fulvous and dark brown.
(Tribe Coppers.)
f^1. Vein 2^3 of fore wings arising at the tip of the cell.
g^1. First joint of middle and hind tarsi not greatly enlarged in male; ground color of upper surface of fore and hind wings the same, or different only in the female.
h^1. Fore tarsi of male jointed; ground color of upper surface of fore and hind wings in the female different.
30. *Chrysophanus.*
h^2. Fore tarsi of male not jointed; ground color of upper surface of all wings the same in the female...31. *Epidemia.*
g^2. First joint of middle and hind tarsi of male twice as stout as rest of tarsus; ground color of all wings above the same in both sexes..32. *Heodes.*
f^2. Vein 2^3 of fore wings arising far beyond the tip of the cell........33. *Feniseca.*
C^2. Of medium or large size, rarely small. Face between eyes as broad as high; eyes not notched next the base of the antennæ.
(Family Typical Butterflies.)
D^1. Antennæ straight; vein 3 of fore wings with three branches; each claw bifid.
(Subfamily Pierids.)
E^1. Antennæ generally very gradually increasing in size to form the club; palpi stout, the last joint short.
(Tribe Yellows or Red-Horns.)

f^1. Club of antennæ cylindrical, broadly rounded at tip.
   g^1. Middle joint of palpi but little longer than broad........34. *Callidryas*.
   g^2. Middle joint of palpi fully twice as long as broad.
      h^1. Vein 2^a of fore wings arising at the tip of the cell ; front margin of fore wings very strongly arched. 35. *Zerene*.
      h^2. Vein 2^a of fore wings arising beyond the tip of the cell ; front margin of fore wings only moderately arched. 36. *Eurymus*.
f^2. Club of antennæ distinctly flattened, the last joint more or less pointed.
   g^1. Club of antennæ very gradually formed and several times longer than broad.
      h^1. Hind femora only about three fifths as long as the middle femora. 37. *Xanthidia*.
      h^2. Hind femora about three fourths as long as the middle femora. 38. *Eurema*.
   g^2. Club of antennæ abruptly formed, hardly more than twice as long as broad. 39. *Nathalis*.
E^2. Antennæ with an abrupt broad flattened club; palpi slender, the last joint about as long as the middle joint.
   F^1. Vein 2^a of fore wings forked near the middle ; middle tibiæ shorter than femora.....................(Tribe Orange Tips.) 40. *Anthocharis*.

F². Vein 2³ of fore wings forked only at the tip; middle tibiæ at least as long as the femora. (Tribe Whites.)
  g¹. Vein 2³ of fore wings arising at or beyond the tip of the cell; fore tibiæ very much shorter than middle tibiæ.
                                41. *Pontia.*
  g² Vein 2³ of fore wings arising distinctly before the tip of the cell; fore and middle tibiæ of equal length.
                                42. *Pieris.*
D². Antennæ more or less arched; vein 3 of fore wings with four branches; each claw simple.
                    (Subfamily Swallow-Tails.)
  e¹. Club of antennæ nearly straight, almost imperceptibly upcurved; tip of abdomen almost reaching emargination of hind wings........43. *Laertias.*
  e². Club of antennæ curved strongly upward throughout; tip of abdomen not nearly reaching emargination of hind wings.
    f¹. Club of antennæ relatively short; hind wings, exclusive of tails, nearly twice as long as broad.
                                44. *Iphiclides.*
    f². Club of antennæ relatively long; hind wings, exclusive of tails, hardly more than half as long again as broad.
      g¹. Fore tibiæ decidedly shorter than the tarsi; tails of hind wings broadened at the end.
        h¹. Vein 4 of hind wings nearly straight; vein 2⁴ of fore wings arising at about

one third the distance from the tip of the cell to the apex of the wing.

i¹. Vein closing the cell of hind wings and connecting veins 2 and 3 not much shorter than the short vein above it ; no transverse stripes on upper side of fore wings.
45. *Jasoniades.*

i². Vein closing the cell of hind wings and connecting veins 2 and 3 less than half as long as the short vein above it; transverse stripes on upper side of fore wings...46. *Euphoeades.*

h². Vein 4 of hind wings strongly sinuous; vein 2¹ of fore wings arising at much less than one third the distance from the tip of the cell to the apex of the wing..............47. *Heraclides.*

g². Fore tibiæ decidedly longer than the tarsi; tails of hind wings not broadened at the end.......48. *Papilio.*

A². Antennæ distant at base, more than half as far apart as the height of the eye, the tip of the club more or less distinctly pointed and recurved; eyes usually overhung next antennæ with a curving pencil of bristly hairs.

(Family Skippers.)

B¹. Recurved part of antennal club nearly or quite as long as the thicker part; abdomen generally shorter than the hind wings.....(Tribe Larger Skippers.)

c¹. Hind wings tailed or distinctly angulate at the tip of vein 4; vein 3¹ arising hardly or no nearer the base of the hind wing

than 2^1; club of antennæ abruptly bent in the middle.
 d^1. Hind wings with a distinct tail or tooth at tip of vein 4............49. *Epargyreus*.
 d^2. Hind wings merely broadly angulate at tip of vein 4..............50. *Thorybes*.
c^2. Hind wings regularly rounded at tip of vein 4 as elsewhere; vein 3^1 arising much nearer the base of the wing than 2^1; club of antennæ curved throughout.
 d^1. Club of antennæ generally ending in a long-drawn point; if not, the antennæ half as long as the fore wing..51. *Thanaos*.
 d^2. Club of antennæ tapering but little on apical half, the tip bluntly pointed, the whole antenna less than half as long as the fore wing.
  e^1. Club of antennæ six or seven times as long as broad, tapering from the middle equally in both directions.
       52. *Pholisora*.
  e^2. Club of antennæ not more than four or five times as long as broad, tapering more rapidly from the middle toward the tip than in the opposite direction.
       53. *Hesperia*.
B^2. Recurved part of antennal club brief as compared with the thicker part, occasionally absent; abdomen reaching to or beyond the outer margin of the hind wing......(Tribe Smaller Skippers.)
 c^1. Club of antennæ with no apical hook.
       54. *Ancyloxipha*.
 c^2. Club of antennæ with a distinct, though sometimes slight, apical hook.

d¹. Hind tarsi shorter than, though sometimes nearly equal in length to, the middle tarsi.
  e¹. Hook of antennal club as long as the width of the club............55. *Atrytone.*
  e². Hook of antennal club shorter, generally much shorter, than the width of the club.
    f¹. Cell of fore wings two thirds as long as the wing...............56. *Erynnis.*
    f². Cell of fore wings only about three fifths as long as the wing.
      g¹. First joint of palpi greatly expanded at tip; middle and hind tibiæ conspicuously spined on the upper surface as elsewhere.....57. *Anthomaster.*
      g². First joint of palpi not expanded at tip; middle and hind tibiæ with no conspicuous spines on upper surface.
        58. *Polites.*
d². Hind tarsi longer than the middle tarsi.
  e¹. Cell of fore wings only three fifths as long as the wing..........59. *Thymelicus.*
  e². Cell of fore wings nearly two thirds as long as the wing..........60. *Limochores.*

KEY TO THE GROUPS, BASED ON THE CATERPILLAR.

A^1. Head and body not separated by a strongly and abruptly strangled neck.
B^1. Body generally covered with spines; when naked or merely covered with pile, either the head is tuberculate, or the last segment ends in a fork, or the body joints are crossed by not more than three creases.
     (Family Brush-footed Butterflies.)
 C^1. Last segment entire, rounded.
  D^1. Body with no spines.
   e^1. Body furnished with a few long fleshy filaments.
     (Subfamily Danaids.) 1. *Anosia*.
   e^2. Body covered with pile only........17. *Anæa*.
  D^2. Body covered with spines.
     (Most of Subfamily Nymphs.)
   E^1. Body uniform, with uniform series of tapering spines.
    F^1. Spines more like tubercles, leathery, not horny, their sides crowded with needles, no one at tip distinguished from the others.
     (Tribe Crescent-Spots.)
     g^1. Body distinctly tapering in front, cross-striped on all but the front segments.

h^1. A tubercle just below the spiracle-line on the third thoracic segment.*

2. *Euphydryas.*

h^2. No tubercle just below the spiracle-line on the third thoracic segment.

3. *Cinclidia.*

g^2. Body scarcely tapering in front, striped longitudinally.

h^1. Tubercles slender, tapering but little, three times as high as broad.

4. *Charidryas.*

h^2 Tubercles stout, conical, less than twice as high as broad......5. *Phyciodes.*

F^2. Spines horny, their sides supporting scattered needles, one at tip crowning the whole.

G^1. No spines along the middle line of the back...............(Tribe Fritillaries.)

h^1. Spines only about half as long as the joints of the body......6. *Brenthis.*

h^2. Spines fully as long as the joints of the body.

i^1. All the spines of upper row equal or subequal and like the rest.

7. *Argynnis.*

i^2. Most of the upper spines of abdominal segments a little longer than the rest, the others nearly equal................8. *Speyeria.*

i^3. Upper spines of first thoracic segment longer than the rest and distinctly enlarged at tip, the others equal.

9. *Euptoieta.*

---

* There is of course no spiracle on this segment; the spiracle-line may be determined by comparing those of the segments next succeeding.

KEY TO THE GROUPS—CATERPILLAR. 47

    G^2. Some spines on the middle line of the back, especially on the seventh or eighth abdominal segment.
        (Tribe Angle-Wings.)
      h^1. Head with no conspicuous spines above.
        i^1. Second abdominal segment with a spine on the middle line of the back.
          j^1. First abdominal segment with a similar spine.........11. *Vanessa.*
          j^2. First abdominal segment with no similar spine...........12. *Aglais.*
        i^2. Second abdominal segment with no spine on middle line above.
          13. *Euvanessa.*
      h^2. Head crowned with prominent spines.
        i^1. Spinules of body spines not arranged in a stellate manner.
          j^1. Spines of thoracic segments with spinules throughout...10. *Junonia.*
          j^2. Spines of thoracic segments with no spinules on basal half. 14. *Eugonia.*
        i^2. Spinules of body spines arranged in a stellate manner.....15. *Polygonia.*
    E^2. Body hunched, with irregularly-developed series of tubercles.
      (Tribe Sovereigns.) 16. *Basilarchia.*
C^2. Last segment bifurcate.
    D^1. Head crowned by a branching appendage.
      18. *Chlorippe.*
    D^2. No coronal spines, or else simple ones on the head.
      (Subfam. Satyrs or Meadow-Browns.)
      e^1. Head with coronal spines or tubercles.
        f^1. Coronal spines slight and inconspicuous.
          19. *Cissia.*
        f^2. Coronal spines nearly as long as the head.

g^1. Head slender and, including the spines, twice as high as broad.
20. *Satyrodes.*
g^2. Head stout and, including the spines, half as high again as broad..21. *Enodia.*
e^2. Head uniformly rounded above 22. *Cercyonis.*
B^2. Body never furnished with spines; the joints crossed by more than three creases, the last joint never forked.
C^1. Body oval and slug-shaped, flattened beneath, rarely almost cylindrical, with very small head.
(Fam. Gossamer-winged Butterflies.)
D^1. Head not one fourth, sometimes not one sixth, the width of the body; dorsal shield behind head wanting or covered with hairs like the parts about it.
(Tribe Blues.)
c^1. Last segment of body broad and greatly flattened ....28. *Everes.*
e^2. Last segment of body comparatively slender and less flattened.....29. *Cyaniris.*
D^2. Head generally at least one third the width of the body; dorsal shield behind head distinct and naked or covered with many fewer hairs than the parts about it.
E^1. Segments of body highest next hind edge, or at least with the hinder slope the more abrupt. Head capable of immense extension.
(Tribe Hair-Streaks.*)

---

* The genera of this group are not sufficiently known to give a key to them.

E². Segments of body highest next front edge, or with the front slope the more abrupt. Head not capable of special extension.
   f¹. Body flattened, covered with short hairs uniformly distributed. 30. *Chrysophanus.* 31. *Epidemia.* 32. *Heodes.*
   f². Body hardly flattened, covered with long hairs arranged in transverse masses. 33. *Feniseca.*

C². Body cylindrical or enlarged in front, with head of ordinary size.
(Family Typical Butterflies.)
  D¹. Back of head descending from summit; body with numerous papillæ and no scent-organs.........(Subfamily Pierids.)
    E¹. Papillæ (supporting hairs) nearly equal in size, or if not, the larger ones are numerous and distinctly arranged in transverse and not longitudinal series on the abdominal segments.
(Tribe Yellows or Red-Horns.)
     f¹. No anterior process on first thoracic segment, above.
      g¹. Papillæ (supporting hairs) elevated, distinctly higher than broad.
       h¹. Papillæ of two sizes, the larger arranged in definite transverse rows.
34. *Callidryas.*
       h². Papillæ of nearly uniform size with no definite transverse arrangement.
        i¹. Largest papillæ on head larger than largest ocelli........37. *Xanthidia.*
        i². Largest papillæ on head smaller than largest ocelli ....38. *Eurema.*

g^2. Papillæ (supporting hairs) mere raised points, not distinctly higher than broad.
    h^1. A shining lenticle just above the spiracle-line on second and third thoracic segments.........35. *Zerene.*
    h^2. No shining lenticle just above the spiracle-line...........36. *Eurymus.*
f^2. A pair of anterior processes on first thoracic segment above.......39. *Nathalis.*
E^2. Papillæ (supporting hairs) of unequal size, the larger arranged in longitudinal as well as sometimes in transverse series on the abdominal segments.
    F^1. Body slender; head much broader than high.
    (Tr. Orange Tips.) 40. *Anthocharis.*
    F^1. Body less slender; head scarcely or not broader than high..(Tribe Whites.)
        g^1. Larger hair-bearing papillæ broader than high................41. *Pontia.*
        g^2. Larger hair-bearing papillæ higher than broad................42. *Pieris.*
D^2. Back of head with no descent from summit; body almost naked, with exceedingly few papillæ and with scent-organs which can be thrust out of the segment behind the head.
(Subfamily Swallow-Tails.)
    e^1. Body with long fleshy filaments on the sides.
43. *Laertias.*
    e^2. Body with no permanent fleshy filaments.
        f^1. Hinder thoracic segments noticeably larger than the next succeeding segments.

g¹. Third thoracic segment with no transverse ridge above.
  h¹. Middle of third thoracic segment without markings ....44. *Iphiclides*.
  h². Middle of third thoracic segment with a pair of eye-like spots.
    i¹. First abdominal segment with no large bright patches above.
              45. *Jasoniades*.
    i². First abdominal segment with a pair of bright patches above, nearly as large as the eye-like spots in front.
              46. *Euphœades*.
g². Third thoracic segment with a transverse dorsal ridge...47. *Heraclides*.
f². Hinder thoracic segments scarcely larger than the succeeding segments.
              48. *Papilio*.
A². Head and body separated by a strongly and abruptly strangled neck..(Family Skippers.)
B¹. Body comparatively stout ; upper half of head as seen from in front rounded or quadrangular...(Tribe Larger Skippers.)
  c¹. Head at least as high as broad, the highest point of each hemisphere lying within the middle line of that hemisphere; dorsal shield obvious.
    d¹. Papillæ of body inconspicuous except from coloring............. 49. *Epargyreus*.
    d². Papillæ of body conspicuous, giving a granulated appearance........50. *Thorybes*.
  c². Head distinctly broader than high, the highest point of each hemisphere at or outside the middle line of that hemi-

    sphere: dorsal shield inconspicuous
    except sometimes at hinder edge.
d¹. Head as seen from in front angulated at upper
    outer corners ; hairs of head simple.
           51. *Thanaos.*
d². Head regularly rounded at upper outer corners;
    hairs of head branching.
  e¹. None of the hairs on abdominal segments
    longer than the shorter sections of
    those segments........52. *Pholisora.*
  e². Among the hairs on abdominal segments are
    some serially arranged which are
    much longer than the sections of
    those segments........53. *Hesperia.*
B². Body very elongated ; upper half of head as seen
    from in front tapering above.
        (Tribe Smaller Skippers.)
  c¹. Head pyramidal, much higher than broad, the
    front facing upward when at rest.
           54. *Ancyloxipha.*
  c². Head more or less rounded, the front facing forward when at rest.

[The further analysis of the genera of Smaller Skippers can hardly be attempted with our present slight information about them.]

KEY TO THE GROUPS, BASED ON THE CHRYSALIS.

A¹. More or less angulated or with projecting shoulders, or if smooth and rounded, then very short and stout, the thoracic spiracle inconspicuous. Not concealed in a cocoon.
B¹. Hanging by the tail only, or else with no hooks at the tail to hang by.
(Family Brush-footed Butterflies.)
C¹. With generally numerous conspicuous prominences.
(Subfamily Nymphs.)
D¹. Head forming a single mass with the thorax.
(Tribe Crescent-Spots.)
e¹. A tubercle on second abdominal segment just above the spiracle-line.
f¹. Tubercles of eighth abdominal segment nearly as prominent as on the preceding segment.....2. *Euphydryas*.
f². No distinct tubercles, but only dark spots on eighth abdominal segment.
3. *Cinclidia*.
e². No tubercle just above spiracle-line on second abdominal segment.
f¹. No distinct ridge uniting tubercles of fourth abdominal segment..4. *Charidryas*.
f². A distinct ridge uniting tubercles of fourth abdominal segment...5. *Phyciodes*.
D². Head projecting independently beyond the thorax.

E^1. Base of wings marked by a pair of tubercles.
F^1. Tail-piece short and stout.
> (Tribe Fritillaries.)
>> g^1. Upper row of tubercles on abdominal segments distinctly unequal in size.
>>> 6. *Brenthis.*
>> g^2. Upper row of tubercles on abdominal segments equal in size.
>>> h^1. Front of head between the smooth crescents tuberculate at the side.
>>>> 7. *Argynnis.* 8. *Speyeria.*
>>> h^2. Front of head between the smooth crescents regularly arched.
>>>> 9. *Euptoieta.*
F^2. Tail-piece long, slender, and tapering.
> (Tribe Angle-Wings.)
>> g^1. Ocellar tubercles blunt and rounded.
>>> h^1. Ridge following upper margin of wings blunt, the dentations rounded.
>>>> 10. *Junonia.*
>>> h^2. Ridge following upper margin of wings sharp, the dentations pointed.
>>>> 11. *Vanessa.*
>> g^2. Ocellar tubercles pointed.
>>> h^1. No tubercle on middle line of second abdominal segment..13. *Euvanessa.*
>>> h^2. A small tubercle on middle line of second abdominal segment.
>>>> i^1. Middle prominence of thorax moderate, almost uniformly tectate.
>>>>> 12. *Aglais.*
>>>> i^2. Middle prominence of thorax large and compressed, at least at tip.
>>>>> j^1. Tubercle just above spiracle-line on

eighth abdominal segment scarcely perceptible..........14. *Eugonia*.

j². Tubercle just above spiracle-line on eighth abdominal segment minute but distinct.........15. *Polygonia*.

E². Base of wings marked by only a single tubercle.

F¹. Middle prominence of thorax very high and strongly compressed.

(Tribe Sovereigns.) 16. *Basilarchia*.

F². Middle prominence of thorax not highly developed.......(Tribe Emperors.)

g¹. Abdomen transversely ridged on the fourth segment, with no longitudinal ridge............. ....17. *Anæa*.

g². Abdomen longitudinally ridged along the middle of the back, with no transverse ridge..........18. *Chlorippe*.

C². With no conspicuous prominences.

D¹. Back of abdomen with a transverse series of tubercles.

(Subfamily Danaids.) 1. *Anosia*.

D². Back of abdomen with no transverse series of tubercles.

(Subfam. Satyrs or Meadow-Browns.)

e¹. Front and lower planes of head forming less than a right angle.

f¹. Abdomen with a pair of distinct longitudinal ridges..............19. *Cissia*.

f². Abdomen with no longitudinal ridges.

g¹. Abdomen beyond tip of wings as long as the wings...........20. *Satyrodes*.

g². Abdomen beyond tip of wings shorter than the wings........21. *Enodia*.

e^2. Front and lower planes of head not forming less than a right angle.

22. *Cercyonis.*

B^2. Fastened around the middle by a silken sling as well as by the tail.

C^1. Body stout, short, and with all projections rounded, the front end broadly rounded.

(Fam. Gossamer-winged Butterflies.)

D^1. Hair-like appendages of the skin cylindrical, pointed, or else stellate at tip.

E^1. These appendages tapering only at tip, the abdomen rarely more than half as long again as broad.

(Tribe Hair-Streaks.*)

f^1. A delicate ridge along middle of thorax.

24. *Incisalia.*

f^2. No distinct ridge along middle of thorax.

g^1. Abdomen much wider than thorax.

h^1. Longest hairs nearly half as long as segments of abdomen..25. *Uranotes.*

h^2. Longest hairs not one fourth the length of abdominal segments. 26. *Mitura.*

g^2. Abdomen scarcely wider than thorax.

27. *Thecla.*

E^2. These appendages tapering throughout or stellate at tip, the abdomen generally almost twice as long as broad.

(Tribe Blues.)

f^1. Body much more than three times as long as broad.... ....28. *Everes.*

f^2. Body much less than three times as long as broad................29. *Cyaniris.*

---

* Chrysalis of Strymon not examined.

KEY TO THE GROUPS—CHRYSALIS. 57

D^2. Hair-like appendages of the skin short, mushroom-shaped..... (Tribe Coppers.*)
 e^1. Abdomen rounded, the last segments not separately protuberant.
  f^1. Only the lower half of ninth abdominal segment sloping forward.
   30. *Chrysophanus.*
  f^2. The whole of ninth abdominal segment sloping forward........32. *Heodes.*
 e^2. Abdomen with irregular surface, the hind segments protruding and expanded.
   33. *Feniseca.*
C^2. Body elongate with angular projections, the front with one or two projecting tubercles.
 D^1. Front end with a single conical projection or rounded prominence.
  (Subfamily Pierids.)
  E^1. Wing-cases distinctly protuberant below the general under surface of the body.
   F^1. The head well distinguished from the frontal projection.
    (Tribe Yellows or Red Horns.)†
    g^1. Ventral protuberance of wings doubling the depth of the body.
     h^1. Fourth abdominal segment with a distinct sharp ridge along the sides.
      34. *Callidryas.*
     h^2. Fourth abdominal segment with no distinct ridge...... 37. *Xanthidea.*
    g^2. Ventral protuberance of wings not doubling the depth of the body.
     h^1. Frontal process slender, acuminate.
      38. *Eurema.*

---

* Chrysalis of Epidemia unknown.
† Excepting Nathalis.

h^2. Frontal process blunt, angulate.
     35. *Zerene.* 36. *Eurymus.*
F^2. Head insensibly merging into the frontal process.
 G^1. Head with no marked projection.
       39. *Nathalis.*
 G^2. Head with an excessively long frontal projection.
   (Tribe Orange Tips.) 40. *Anthocharis.*
E^2. Wing-cases scarcely protuberant below the general under surface of the body.
      (Tribe Whites.)
 f^1. Frontal process stout, no longer than broad.
      41. *Pontia.*
 f^2. Frontal process slender, very much longer than broad............ 42. *Pieris.*
D^2. Front end with two projecting tubercles.
    (Subfamily Swallow-Tails.)
 e^1. Surface of body except the large projections tolerably smooth.
  f^1. Distinct ridges along the sides of abdomen above.
   g^1. Abdomen greatly expanded next the base.
      43. *Laertias.*
   g^2. Abdomen gently enlarged in the middle.
      44. *Iphiclides.*
  f^2. No ridges along sides of abdomen above.
      46. *Euphœades.*
 e^2. Surface of body very much roughened.
  f^1. Under surface of body, as seen from the side, hardly bent... 45. *Jasoniades.*
  f^2. Under surface of body, as seen from the side, strongly bent.
   g^1. Base of antennæ with a distinct tubercle.
      47. *Heraclides.*

KEY TO THE GROUPS—CHRYSALIS.

    g^2. Base of antennæ with no tubercle.
            48. *Papilio*.
A^2. Smooth and rounded, elongate, the thoracic spiracle conspicuous. Concealed in a cocoon..........(Family Skippers.)
 B^1. Tongue-case not free, not extending beyond the wings.....(Tribe Larger Skippers.)
  c^1. Abdomen exclusive of tail-piece no longer than the rest of the body.
   d^1. Thoracic spiracle with no posterior elevated flaring lip........ 49. *Epargyreus*.
   d^2. Thoracic spiracle with a posterior elevated flaring lip.............50. *Thorybes*.
  c^2. Abdomen exclusive of tail-piece longer than the rest of the body.
   d^1. Hinder lip of thoracic spiracle scarcely raised, not flaring...........51. *Thanaos*.
   d^2. Hinder lip of thoracic spiracle much elevated, flaring, fluted.
    e^1. The hinder equal part of tail-piece, seen from above, scarcely longer than broad.
            52. *Pholisora*.
    e^2. The hinder equal part of tail-piece, seen from above, twice as long as broad.
            53. *Hesperia*.
 B^2. Tongue-case free at tip, extending beyond, sometimes much beyond, the wings.
         (Tribe Smaller Skippers.)

[The genera of Smaller Skippers are too little known to separate them by their chrysalids.]

NOMENCLATURE OF THE PARTS OF THE WING.

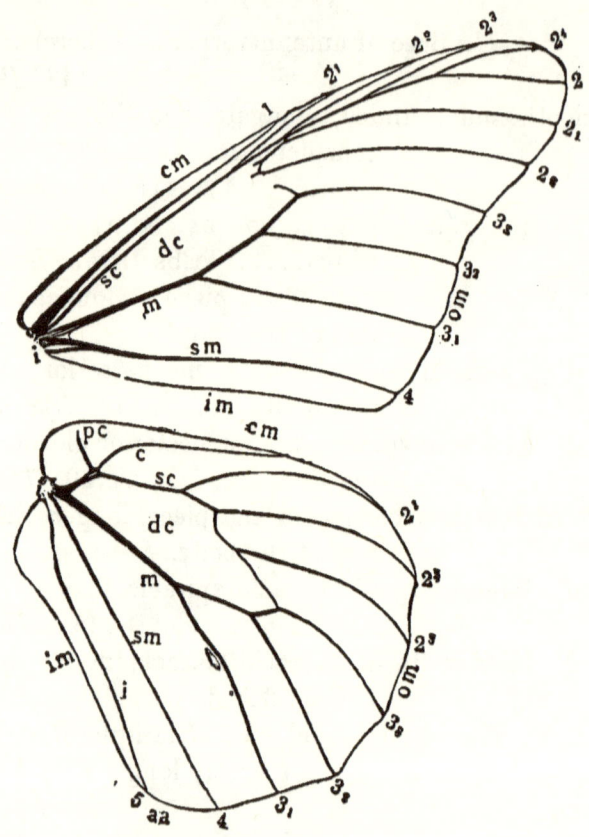

NEURATION OF ANOSIA PLEXIPPUS.

cm .....costal margin.
om......outer margin.
im.......inner margin.
dc.......discoidal cell.
aa......anal angle.

pc (0) .....precostal vein.
c (1)..... costal vein.
sc (2)......subcostal vein.
m (3)......median vein.
sm (4).....submedian vein
i (5)..... internal vein.

NOMENCLATURE OF THE PARTS OF THE WING.

The veins may for conciseness, as in our "Key to the Groups," be numbered from above downward from 0 to 5 as in their explanation above, and their branches may be further indicated by adding to the number one which shall designate whether it is the first branch, second branch, etc., and also whether it is thrown off from the upper or under edge. Thus the branches striking the margin of the fore wing in the above figure, beginning above, would have this consecutive designation: $1, 2^1, 2^2, 2^3, 2^4, 2, 2_1, 2_2, 3_3, 3_2, 3_1, 4$ (the internal running into the submedian); while those of the hind wing (including the postcostal, which does not quite reach the margin) would be: $0, 1, 2^1, 2^2, 2^3, 3_3, 3_2, 3_1, 4, 5$. In this way equivalent nervules of the two wings, or of the same wings in different butterflies, would have a similar symbol.

THE COMMONER BUTTERFLIES

OF THE

NORTHERN UNITED STATES AND CANADA.

FAMILY BRUSH-FOOTED BUTTERFLIES.

SUBFAMILY DANAIDS.

1. GENUS ANOSIA.

ANÒSIA PLEXÍPPUS—THE MONARCH OR MILK-WEED BUTTERFLY.

(Danais archippus, Danais erippus.)

Butterfly.—Wings above and fore wings beneath rather light tawny brown, the veins margined with black, and the wings broadly margined with the same enlivened by a double row of small whitish spots; besides, all the apex of the fore wings is more or less black, but contains two or three dashes of obscure tawny and, just beyond the cell, a couple of oblique series of large buff-tawny spots, those nearest the front margin smaller, elongate, and white. Beneath, the ground color of the hind wings is buff, and the black veins are edged with some whitish scales. The male is distinguished by a conspicuous thickened black patch (really a pocket containing special scales) next one of the veins near the middle of the hind wings. Expanse 4 inches.

Caterpillar.—Head smooth and rounded, yellow, conspicuously banded with black. Body cylindrical, tapering a little in front, naked, but with two pairs of long and very slender black thread-like filaments, one pair, the longer, on the second thoracic, the other on the eighth abdominal, segment. The body is white with numerous slender black and yellow, and especially black, transverse stripes, repeated with considerable regularity on each of the segments, so that there are nowhere any broad patches of color. Length nearly 2 inches.

Chrysalis.—Pea-green. Stout and not elongated, largest in the middle of the abdomen, where it is transversely ridged; elsewhere it is smooth and rounded, with no striking prominences, but with little conical projections at most of the elevated points like those which half encircle the body at the abdominal ridge, all of a golden color except the latter, which are situated in a tri-colored band, black in front, nacreous in the middle (these dividing the points between them), and gilt behind. Length more than 1 inch.

We begin with one of the most interesting of our butterflies, about which a volume might be written, but of which we have still much to learn. It is found in the summertime over almost the entire continent, certainly as far north as into the Dominion of Canada; and yet it is probable that it does not exist in the winter further north than the Gulf States. It has extraordinary powers of flight, more so than any known butterfly, and every autumn when abundant (after first collecting in vast flocks or bevies of hundreds of thousands, changing the color of the trees or shrubs on which it alights for the night) migrates southward in streams, like our migrating birds. After passing the winter on the wing, without so far as known hibernating in torpidity, it leaves its winter quarters in the extreme south with the opening spring and flies northward, not in flocks or streams, but singly. The females lay their eggs when they are ripe wherever they may chance to be, some flying even as far as southern New York and Minnesota before concluding their life-duties. The caterpillars born from these eggs develop into butterflies, many of which again fly northward before they lay their eggs; while the butterflies developing from these last do not lay eggs the same season (unless possibly in the warmer south), but migrate southward at the end of the season, to return again the next spring. North, therefore, of the farthest points to which the wintering butterflies have journeyed in the spring, there appears to be but one brood a year, south of it two, and in the extreme south possibly more.

As a further proof of the transcendent powers of flight of this butterfly, it may be mentioned that it has been seen at sea five hundred miles from land and has within thirty years spread over nearly all the islands of the Pacific and even to Australia and Java. Undoubtedly carried in the first place by trading or other vessels to the Hawaiian Islands and thence to Micronesia, it has unquestionably *flown* from island to island many hundreds of miles apart. It has also appeared at various times in different places on the sea-coast of Europe, here also probably transported accidentally by vessel. In 1885, for instance, no less than nine specimens were captured in four different counties of England, and in 1886 it was reported at different points from England to Gibraltar.

The egg is long oval in shape, with over twenty low upright ridges and many cross lines, is of a pale green color, and is laid singly on the food-plant of the caterpillar (various kinds of milk-weed, especially the commonest kind, *Asclepias cornuti*) and usually upon the under surface of the tender upturned apical leaves near the middle. It hatches in about four days, the caterpillar feeds quite exposed upon the leaves, generally resting, however, upon the under surface, and takes two or three weeks to grow to its full size. In New England the eggs are usually laid during July, and belated caterpillars may be found even in September. The chrysalis hangs from nine to fifteen days.

But the chief interest attaching to this butterfly is that

it belongs to a favored race, as, like all the members of its tribe, it is protected from its natural enemies among the birds by some nauseous peculiarities. The males can protrude from the end of the abdomen on either side a bunch or brush of hairs which may be the means of producing an offensive smell; but besides this the whole body of both sexes seems to have a rank odor, and its protection is the cause of its unconscious mimicry by another of our butterflies, *Basilarchia archippus*. It is the best example of mimicry known in North America.

The subfamily of Heliconians is represented in the southern part of our district by the genus **Agraulis**, with one species, *A. vanillae*, a southern species which has occasionally been taken as far north as Pennsylvania.

SUBFAMILY NYMPHS.

TRIBE CRESCENT-SPOTS.

2. GENUS EUPHYDRYAS.

EUPHÝDRYAS PHÀETON—THE BALTIMORE.

(Melitaea phaeton.)

Butterfly.—Wings black, marked with red and pale straw-yellow, the markings larger on the under than on the upper surface; the red is confined to two or three spots (more below) near the base of each wing and to a broad outer margin, divided by the black veins; the yellow mostly to four parallel series (two on the upper surface of hind wings) of small round or squarish spots (the outer row lunulate) between the veins in the outer half of the wing, before the marginal band. Expanse of male 2 inches; of female 2½ inches.

Caterpillar.—Head black with low conical summits. Body spined, dark orange transversely ringed with black lines, the thoracic segments tapering, mostly black; spines bluish black, about as long as the segments with numerous long black bristles, set on papillae; there is a dorsal series, two others on each side equally dividing the space between that and the spiracles, and one other below the spiracles, including one on the third thoracic segment; a row of smaller spines, two to a segment, occurs at the base of the prolegs. Length 1 inch.

Chrysalis.—Very pale bluish white, marked with velvety black and pale orange; little conical orange tubercles mark the position of the spines of the caterpillar, black dots or small dashes are sprinkled over the body especially on the abdomen, and larger dashes divided by orange nervules cross the middle of the wings in a continuous series. Legs orange marked with black. Tubercles of eighth abdominal segment distinct. Length nearly $\frac{3}{4}$ inch.

The eggs are largest below, taper above to a very broad and depressed summit, the sides vertically ribbed on upper half, at first yellow, afterwards purplish; they are laid in large irregular clusters, several layers deep, upon the under surface of a leaf of the food-plant, and hatch in about twenty days. During the season in which they are born the caterpillars feed in society, living in a web with which they line and envelop their food-plant, the snake-head, *Chelone glabra*, and less commonly other Scrophulariaceous plants. After moulting three times, which the caterpillars do under and within their webs, the whole colony hibernates within the web, made more dense for the purpose, which, contracting as the winter dries the foliage, becomes a compact rounded mass as large as an egg, filled with caterpillars, cast skins, and filth. In the spring the caterpillars make their way out, disperse, and no longer construct webs but feed openly, frequently choosing other food-plants, Lonicera or Viburnum, Caprifoliaceous plants. The chrysalis hangs from fourteen to eighteen days. The butterfly is extremely local, often confining its wanderings to an acre of ground, and is only found near or in swampy places; it flies heavily—indeed it is our most sluggish butterfly—and is single-brooded, appearing early in June and flying for more than a month.

3. Genus Cinclidia.

CINCLÍDIA HARRÍSII—HARRIS'S BUTTERFLY.

(Melitaea harrisii, Phyciodes harrisii.)

Butterfly.—Upper surface of wings nearly black, the fore wings with a broad sinuous band of dull orange across the middle broken by the black veins, followed outwardly by a sinuous row of similar unequal spots and inwardly by a few irregular orange spots ; hind wings with most of the disk dull orange, begrimed with black and cut by black veins. Under surface brownish orange, the veins mostly black, marked with usually black-edged white spots, conspicuous on the hind wings where the median spots are sordid cut by a black line, the subbasal and lunular subapical spots shining. Expanse 1¾ inches.

Caterpillar.—Head shining black, summits tuberculate and low conical. Body spined, tapering on the thoracic segments, deep orange with a black dorsal line, and ringed narrowly with black stripes throughout ; spines jet-black, a little shorter than the segments, covered with black needles set on papillae ; they are arranged as in Euphydryas excepting that there is no spine on the third thoracic segment in the row just below the spiracles. Length nearly 1 inch.

Chrysalis.—Snow-white, marked much as in *Euphydryas phaeton*, but with the darker markings mostly confined to edgings of the orange tubercles. Legs white tipped with black. No distinct tubercles on the eighth abdominal segment, but their place marked by spots. Length ½ inch.

The eggs, which are shaped as in Euphydryas but with a smaller summit, are pale lemon-yellow and are laid in patches of twenty or more in a closely-crowded single layer on the under side of a leaf of the food-plant; their period in unknown. So far as known, the caterpillars have but a single food-plant, *Aster* (*Doellingeria*) *umbellatus*. They first eat the parenchyma of the under surface of the leaf on which they are born and then move in company down the plant, devouring the parenchyma of each surface of every leaf as they go, covering everything with a thin web, beneath and upon which they live until the end of the season, their nests resembling those of Euphydryas but less dense. Early in September and after two or three moults they desert these nests and hibernate in crannies, probably to some degree in company; for in the early spring they may be found again in loose companies, but living openly, often four or five on a single leaf of their food-plants and in close vicinity to their birthplace. The caterpillars change to chrysalis at the end of May or early in June and hang from ten to sixteen, usually about thirteen, days. The butterfly is extremely local, occurring only in the immediate vicinity of places where the food-plant grows; but not always there, for the butterfly hardly occurs south of lat. 42° or west of Wisconsin, while Doellingeria extends to Georgia and Arkansas. It is single-brooded, appearing upon the wing about the middle of June and flying throughout July.

4. Genus Charidryas.

CHARÍDRYAS NYCTEIS—THE SILVER CRESCENT.

(Melitaea nycteis, Phyciodes nycteis.)

Butterfly.—Upper surface of wings pale orange fulvous, marked with black; fore wings with outer border very broadly margined with black, especially above, where it nearly reaches a broad bar descending from the costa to the middle of the wing; base and cell with a tangle of black lines; hind wings mostly black with an exceedingly broad subequal transverse fulvous belt, broken in the middle by a brown stripe and with a row of round spots in outer half. Under surface of fore wings much like upper (but washed out) excepting for varied light markings near apex; hind wings pale buff marked with dark brown, the veins brown, dull silvery spots next the base and one or two on the costal and apical margins, on the latter in the middle of a broad brown field. Expanse 1¾ inches.

Caterpillar.—Head shining black, rounded on summits. Body spined, scarcely tapering on thoracic segments, velvety black above with a dull orange stigmatal band; spines black or blackish, slender, at least three times as high as broad, arranged much as in Euphydryas. Length nearly 1 inch.

Chrysalis.—"Some are light-colored, nearly white, with delicate blackish spots and fine streaks of brown over the surface; others are almost wholly black, while others again are between the two extremes" (Edwards). It closely resembles that of *Cinclidia harrisii*, from which it may be distinguished by having no suprastigmatal tubercle on the second abdominal segment, and by the wing spots hardly forming a definite band. Length $\frac{1}{2}$ inch.

The eggs, the sides of which are ribbed above, pitted in the middle, and smooth below, are pale green and are laid on the under surface of a leaf of the food-plant in clusters of from a few up to a hundred, side by side in regular rows; they hatch in from nine to fourteen days. The caterpillars feed on various Composite plants, particularly sunflower and Actinomeris; when young they are gregarious and feed on the parenchyma of the leaf; later they eat the whole leaf, but at no time do they spin a web for concealment or protection; they hibernate when partly grown, doubtless in crevices, and separate in spring, feeding singly. The chrysalis hangs from ten to fifteen days. The butterfly is not at all local and is far more common in the West than in the East, where it has not been recognized east of the middle of Maine. It appears to be single-brooded in the North, flying in the latter half of June and in July; but according to observations in West Virginia and Missouri it appears to be there partly single- and partly double-brooded, a first generation appearing in May and a second, partial generation in July, some of the caterpillars from the May butterflies going into early hibernation, others passing forward to form the second generation.

Another species of this genus is *C. ismeria*, which is a southern form, but in the West occurs as far north as Colorado and Montana and has even been reported from Brandon, Manitoba.

5. GENUS PHYCIODES.

PHYCIÒDES THÀROS—THE PEARL CRESCENT.

(Melitaea tharos, Melitaea marcia, Melitaea pharos.)

Butterfly.—Wings dull orange, heavily marked with blackish brown, the markings heavier in the female and found on the upper surface principally in a broad outer margin, a broad divided bar across the middle of the fore wings, and a mesh of lines, confused in the female, at the base of the wings ; a preapical series of dots on the hind wings. On the under surface the dark markings of the fore wings are mostly confined to irregular patches at the middle of the costal and at the middle and just before the tip of the inner border; the hind wings are ochraceous with a transverse median tracery of lunulate cinnamon lines, and a large brown cloud on the hind margin ; the preapical dots of the upper surface are repeated. Expanse $1\frac{1}{2}$ inches.

Caterpillar.—Head shining bronze, marked with white, rounded on summits. Body spined, scarcely tapering on thoracic segments, blackish, dotted above with yellow, with a black dorsal stripe (often wanting), a yellow line in the middle of the sides, and a yellow band just beneath the spiracles; spines mostly yellowish, stout, less than twice as high as broad, arranged much as in Euphydryas. Length $\frac{3}{4}$ inch.

Chrysalis.—Grayish white, the effect of brownish creases on a white ground, darker on the abdomen, where there is a dull band below the spiracles ; no band on the wings. Length $\frac{5}{8}$ inch.

The eggs, which taper so that the summit is only half as broad as the base and are ribbed above on the sides, are light yellow-green and are laid in clusters of from twenty to two hundred on the under side of the leaves of the food-plant, crowded together, sometimes in one layer, at others in several ; they hatch in from five to ten days. The caterpillars feed on asters, but their proper food-plant appears to be only *Aster novae angliae*. They feed in company, devouring at first only the parenchyma of the under surface of the leaf, later in life the entire leaf, spinning no web at any time. The caterpillars of the latest brood become lethargic after the second or third moult and then hibernate. The chrysalis hangs for an uncertain period, generally from six to thirteen days, sometimes prolonged to a month. The butterfly flies slowly and for short distances only; it is everywhere abundant in open places and is single- or double-brooded according to locality, triple- or even quadruple-brooded further south. In New England it is double-brooded, the first brood appearing in the latter half of May and flying until the end of the first week in July; the second brood appears about the middle of July and may be found even to October, there being great irregularity in the development of different caterpillars, among which there is sometimes a certain amount of temporary lethargy. The full accounts of the behavior of the caterpillars of this species given by Mr. W. H. Edwards are well worthy of close attention. The species is dimorphic, the butterflies of the first brood (wherever there are more than two) differing from those of the later in having more accentuated markings.

A second species of this genus, *P. batesii*, has been taken sparingly east of the Appalachians; and a third, *P. gorgone*, an extreme southern species, has been recorded from Kansas.

TRIBE FRITILLARIES.

6. Genus Brenthis.

BRÉNTHIS BELLÒNA—THE MEADOW FRITILLARY.

(Argynnis bellona.)

Butterfly.—Upper surface of wings fulvous, heavily marked with black; on most of the basal half or more, bounded by an angulate dentate outer line, the black predominates, touched with fulvous dashes; outer margin bordered with black reduced to small T-shaped spots on the hind wings, preceded by two rows of spots, the inner circular and crossing the middle of the fulvous field. On the under side the fore wings are fulvous heavily blotched with black excepting on the outer fourth, where there are cinnamoneous clouds; hind wings cinnamoneous fulvous on the basal half, one or the other tint predominating in large spots, traversed by brown lines, the outer half purplish brown, obscurely clouded and marked with brown. Expanse nearly 2 inches.

Caterpillar.—Head shining blackish green, the summits rounded. Body spined, purplish black, mottled with yellowish and with a velvety-black broken lateral stripe ; spines leathery, dull luteous tipped with fulvous, all of nearly the same size. Length nearly 1 inch.

Chrysalis.—Dark yellowish brown, resulting from brown creases on a yellowish-brown ground ; laterodorsal tubercles of abdomen (very prominent on third segment) constricted before the tip, those of first and second segments of equal size. Length more than ½ inch.

The eggs, which are tall sugar-loaf-shaped with twenty or more prominent vertical ribs, are dull olive-yellow and are probably laid singly on the food-plant ; one observer says he has seen the female drop her eggs loosely while hovering in the air ; they hatch in from five to nine days. The caterpillars feed singly and openly upon violets, but only at night, making no web and concealing themselves about the roots of the herbage by day. Winter is passed by the caterpillars when half grown. The chrysalis hangs for about a week. The butterfly is most commonly found about wet meadows and bogs, and is a northern species, hardly found south of lat. 41°; it has a moderately rapid but low zigzag flight. There are three broods annually: the first appears about the middle of May and fresh specimens continue to emerge throughout June ; the eggs, however, appear not to be laid until the middle of June and may be laid all through the rest of the month and July, for the butterfly is very long-lived; the second brood appears about the middle of July before the first brood has disappeared and continues on the wing into September; the third brood appears late in August and continues up to the time of frosts.

There are some strange anomalies about the development

of this butterfly. It would appear that in the first brood of butterflies, and sometimes but not always in the second, the eggs are not developed in the bodies of the females so as to be ready to lay until the butterfly has been on the wing two or three weeks; while in part of the second and all of the third brood the eggs are fully developed as soon as the butterflies emerge from the chrysalis, or at any rate in a day or two. So, too, the behavior of the caterpillars is very different, at least in the second brood, some feeding regularly and passing forward to form the chrysalids from which the butterflies of the third brood emerge; others becoming lethargic in midsummer, when half grown, and passing into premature hibernation curled up in crannies. As the caterpillars from the eggs of the final brood of butterflies probably hibernate before moulting at all, the spring opens with caterpillars of different stages of growth and of different generations of the preceding year, which passing on to chrysalis combine to make the first long-drawn-out brood of butterflies. Whether any of the caterpillars of the first brood behave in this way (so that the spring brood of butterflies shall be made up of parts of *all* the generations of the preceding season) is not yet determined, but it seems probable from the irregularity and long continuance of the second brood of butterflies.

BRÉNTHIS MYRÌNA—THE SILVER-BORDERED FRITILLARY.

(Argynnis myrina.)

Butterfly.—Upper surface of wings fulvous marked with black; the markings consist principally of an outer margin inwardly dentate and enclosing fulvous dots, a curving series of round spots beyond middle of outer half of wing, and across the base and middle a coarse and irregular mesh of subcontinuous dashes. Under surface of fore wings fulvous with black markings feebly repeated, a cinnamoneous cloud at apex and apical silvery spots; of hind wings mixed cinnamoneous and ochraceous, with two transverse series of silvery spots, besides those at base and apex. Expanse 1¾ inches.

Caterpillar.—Head dark metallic green, the summits rounded. Body spined, mottled with dark green, purple, and luteous; spines leathery, blackish fuscous or partly luteous, those on the back of the first thoracic segment several times longer than the others. Length ¾ inch.

Chrysalis.—Dark luteous, the abdomen darker, the whole marked with fuscous; laterodorsal tubercles of abdomen (very prominent on the third segment) uniformly conical, those of first segment smaller than those of second. Length ½ inch.

The eggs, which are tall sugar-loaf-shaped, with sixteen or seventeen prominent vertical ribs, are olivaceous yellow and are laid singly on the leaves or stems of the food-plant or on immediately adjoining vegetation; also, according to some observers, dropped loosely on the wing; they hatch in from six to ten, sometimes fourteen, days. The caterpillars feed by night upon violets, and hide by day, and are very quick in their movements and easily disturbed. The chrysalis hangs from seven to eleven days. The haunts and flight of the butterfly are the same as those of *B. bellona* and its life-history probably identical; certainly it passes the winter in the caterpillar state, both just from the egg and half grown, but the lethargic features noticed in the preceding species have not been observed, though they probably occur, in this; the butterfly, however, is a few days later than *B. bellona* in appearing in its successive broods in a given locality.

Three other species of Brenthis occur in the northern parts of our district, two in the high north, *B. chariclea* and *B. freija*, both of them circumpolar insects, sometimes taken in Canada not far from our border; and *B. montinus*, known only from the subalpine districts of the White Mountains of New Hampshire, and thought by some to be merely a variety of *B. chariclea*.

7. GENUS ARGYNNIS.

ARGÝNNIS ATLÁNTIS—THE MOUNTAIN SILVER-SPOT.

Butterfly.—Upper surface of wings orange fulvous, with well-defined black markings. These consist, in all our species of Argynnis, in the fore wings, of three sinuate bars across the outer part of the cell besides a straight and a sinuate bar at the tip, a more or less disconnected zigzag band across the middle of the wing, and a series of rounded spots on the middle of its outer half, besides a submarginal series of sagittate spots on a dusky border. On the under surface the design of the fore wings is a vague repetition of the upper markings, while the hind wings have submarginal, extramesial, intramesial, and prebasal series of very large silvery spots, those of the outer series usually the larger. The peculiarities of each species are seen principally on the under surface of the hind wings, which in the present species is distinguished by the depth and griminess of the basal tint and by the width of the buff belt between the two outer rows of silver spots, which is intermediate in this particular between *A. aphrodite* and *A. cybele*. Expanse 2½ inches.

Caterpillar.—Head dark. Body spinous, dark velvety purple above, scarcely paler beneath; spines corneous, livid at base, the spinules nearly half as long as the spines. Length 1¼ inches.

Chrysalis.—Chestnut-brown irrorate with black, basal segments of abdomen unicolorous; dorsal and ventral surfaces of front part of body set at an angle of about 50°. Length ⅞ inch.

The eggs, which are short sugar-loaf-shaped, as high as broad, with twelve to fourteen vertical ribs and honey-yellow, are laid singly on the food-plant and hatch in about a fortnight. The caterpillars go into winter quarters immediately after emerging from the egg without tasting of vegetable food, awake early in the spring, and feed singly and by night upon violets, hiding in crevices by day. The chrysalids are found attached to the under side of logs lying on the ground and in similar places; their period is unknown. The butterfly is wilder than the succeeding species of the genus and is a more northern form, being limited southwardly by about the annual isotherm of 45° F.

It is single-brooded, appearing about the middle of June, but not becoming common until the middle of August, and is still on the wing in September; although the males appear some time before the females, the latter may be found long before they are ready to lay their eggs, which is not until the latter half of August. The males have a very perceptible odor of sandal-wood.

ARGÝNNIS APHRODÌTE—THE SILVER-SPOT FRITILLARY.

Butterfly.—The ground color of the under surface of the hind wings is a pure cinnamoneous, and the buff band between the two outer rows of silver spots is very narrow, narrower than the outermost brown margin, and at its extremities often disappears. Expanse 3 inches.

Caterpillar.—Head black, reddish yellow behind. Body spined, blackish brown, with a velvet-black spot at base of each spine, not so dark beneath; spines corneous, black, some reddish yellow at base. Length fully 1½ inches.

Chrysalis.—Livid brown and blackish, less coarsely rugose, and with less prominent tubercles than in *A. cybele*, the basal segments of the abdomen bicolored. Length nearly 1 inch.

The eggs, which are short sugar-loaf-shaped, as high as broad, with sixteen to nineteen vertical ribs and honey-yellow, are laid singly on the food-plant and hatch in a fortnight. After devouring their egg-shells, the caterpillars move actively about as if searching for winter quarters, utterly declining all vegetable food. After hibernation they feed by night on all kinds of wild violets, and during the day lie concealed on the ground under chips and stones; they are very active. The chrysalis hangs from seventeen to twenty days. The butterfly is very fond of the blossoms of the thistle, and when feeding can readily be taken with the fingers. Though a more northern butterfly than *A. cybele*, it is more southern than *A. atlantis* and more eastern than *A. alcestis*. It is found throughout New England, excepting in the heart of the White Mountains. It is single-brooded and a little later in appearance than its companion-species, first appearing about the beginning of July; the butterflies are seldom abundant before the end of the first week in July, and disappear by the middle of September; the eggs are not laid, apparently, before the middle of August. The males have no perceptible odor.

ARGÝNNIS ALCÉSTIS—THE RUDDY SILVER-SPOT.

Buttterfly.—The ground color of the under surface of the hind wings is a nearly uniform and pure deep cinnamoneous, with no distinct band of buff between the outer rows of silvery spots. Expanse 3 inches.
Caterpillar.—Head black, yellowish behind. Body spinous, velvety black; spines corneous, black above the yellowish base. Length 1⅝ inches.
Chrysalis.—Red, brown, or drab, irregularly mottled and creased with black; abdominal segments drab, edged in front with black. Length 1 inch.

The eggs, which are short sugar-loaf-shaped, much higher than broad, with about eighteen vertical ribs, are presumably laid on the food-plant and hatch in from twenty-five to thirty days. Nearly all the caterpillars, after devouring their egg-shells, go at once into hibernation, but some have been known (in captivity, in a region south of their native home) to feed and moult once or twice before winter; they feed readily on violets. The chrysalis hangs for three weeks or more. The butterfly is fond of the open country and is found only in the West, occurring in the Mississippi Valley from Michigan to Montana north of lat. 40°. Its seasons are all similar to those of our eastern species of Argynnis. The male has been credited with no odor.

ARGÝNNIS CÝBELE—THE GREAT SPANGLED FRITILLARY.

Butterfly.—The ground color of the under surface of the hind wings is rather dull cinnamoneous, more or less sprinkled with buff, and the buff band between the two outer rows of silver spots is very broad, broader than the outermost brown border, and extends from margin to margin. Expanse fully 3 inches.
Caterpillar.—Head dull black, castaneous behind. Body spined, dull black, the more exposed parts somewhat velvety; spines corneous, shining blackish castaneous, the base of many dull orange luteous. Length 1⅜ inches.
Chrysalis.—Dark brown, creased and mottled with drab or reddish brown, or almost wholly dead-leaf brown, more coarsely rugulose and with more prominent tubercles than in *A. aphrodite*, the basal segments of abdomen unicolorous. Length more than 1 inch.

The eggs, which are short sugar-loaf-shaped, higher than broad, with sixteen to eighteen vertical ribs, and honey-yellow, are laid singly on the food-plant, and also, according to some observers, loosely dropped by the mother while poising in the air; they hatch in about fifteen days. The caterpillars go at once into hibernation, and become full fed on violets during the next June. When about to pupate, the caterpillar seeks the under surface of stones and of bark lying on the ground, and the chrysalis hangs from fourteen to twenty-four days. The butterflies are found in open fields and are single-brooded, the earliest appearing the last of June and continuing to emerge from the chrysalis until at least the middle of July; they remain on the wing until the middle of September or later; although pairing by the end of July, the earliest females not appearing until the beginning of that month, eggs are hardly laid before the middle of August. Further south, according to W. H. Edwards, the butterflies appear at the end of May, but by the first of July have all disappeared, a fresh brood appearing about the middle of August; yet he has never been able to get butterflies of this first brood to lay eggs, nor has he found mature eggs in the bodies of females at that season. The male has no perceptible odor.

8. Genus Speyeria.

SPEYERIA IDALIA—THE REGAL FRITILLARY.

(Argynnis idalia.)

Butterfly.—Upper surface of fore wings brilliant orange, marked with black, much after the pattern of Argynnis; of hind wings purplish black, with an extramesial bent series of cream-colored roundish spots and a submarginal series of similar spots, cream-colored in the female, orange in the male. Under surface of fore wings as in Argynnis, of hind wings dark olivaceous, heavily marked, Argynnis-fashion, with series of large silvery spots, edged, especially on the basal side, with black. Expanse 3½–4 inches.

Caterpillar.—Head black below, reddish above. Body spinous, velvety black, heavily banded and striped with ochrey yellow or reddish; spines corneous, mostly yellowish, the spinules black. Length 1¾ inches.

Chrysalis.—Brown, tinged with pink and marked with black in rather small spots, scattered over the thorax and wings and in front of, sometimes including, the tubercles. Length more than 1 inch.

The eggs, which are short sugar-loaf-shaped, broader than high, tapering rapidly, with sixteen to eighteen vertical ribs and pale green, are laid singly on the food-plant, probably on the under side of the leaves; they hatch in about thirty days. The caterpillars at once hibernate after devouring their egg-shells, or possibly some remain in the egg all winter. The remainder of the life-history transpires the next season, the caterpillar feeding upon violets (and Compositae ?), the chrysalis hanging (in the single instance recorded, in West Virginia) seventeen days. The butterfly is somewhat local and is found in open breezy places, occurring only in a relatively narrow belt across the country, following the annual isotherm of 50° F.; it flies low and with no great rapidity, settling suddenly, and is single-brooded, the males appearing at the very end of June or early in July, the females about ten days later, and both continuing on the wing until near the end of September, fresh specimens coming from the chrysalis until after the middle of August, indicating probably some lethargy in the caterpillars. The eggs are not laid until the last of August and usually not until September. This is one of our showiest butterflies and the male has a slight musky odor.

9. Genus Euptoieta.

EUPTOIÈTA CLAÙDIA—THE VARIEGATED FRITILLARY.

(Argynnis columbina.)

Butterfly.—Upper surface of wings pale fulvous, darker in the basal half, with an irregular, transverse, black mesial line, darker, broader, and much more abruptly zigzag on the fore than on the hind wing, and a pair of extramesial, more or less wavy brown lines enclosing between them a series of round blackish spots. Under surface of fore wings much like the upper, with the addition of a large apical clouded patch of gray and brown, obliquely divided; of hind wings dark yellowish brown with the markings of the upper surface obscurely repeated and overlaid by hoary patches and streaks, especially forming a marginal and a broad extramesial band, in both more intense in tint toward the costal margin. Expanse more than 2 to nearly 3 inches.

Caterpillar.—Head blackish, orange above. Body spinous, very variable in color but generally of some glistening shade of reddish orange, twice longitudinally banded on each side with black, enclosing or partly enclosing squarish white spots. Length 1¼ inches.

Chrysalis.—Silvery white, dotted and blotched with black; wings much blotched with black; tubercles gilt, but sometimes silvery behind, nearly encircled with black. Length ¾ inch.

The eggs, which are short sugar-loaf-shaped, with from thirty to forty vertical ribs and pale green, are laid singly on the food-plant and hatch in from five to twelve days.

The caterpillar feeds on a considerable variety of polypetalous plants, but particularly on Passiflora and Sedum; it feeds readily on violets and has been known to be injurious to the garden pansy; it probably feeds only by night. The chrysalis hangs for about eleven days. The butterfly frequents open fields and is a southern form, though occurring farther north in the Mississippi Valley than in the East; it is rarely found in southern New England and perhaps does not winter there. It is apparently triple-brooded; the last brood is the most numerous and appears so late that, taking into account the appearance of butterflies very early in the spring, it seems probable either that the butterfly itself hibernates or else that some of the autumn chrysalids continue over the winter, or both; but it is not unlikely also that caterpillars from eggs laid late in the season may hibernate as soon as hatched or when partly grown. It is only by further careful observation and experiment in the Middle and Southern States that the life-history of this butterfly can be determined. The inequality of the broods would indicate lethargic tendencies in midsummer caterpillars.

The genus **Semnopsyche** (*S. diana*) also occurs in the southernmost part of our district.

TRIBE ANGLE-WINGS.

10. GENUS JUNONIA.

JUNONIA CŒNIA—THE BUCKEYE.

(Vanessa coenia, Junonia lavinia).

Butterfly.—Upper surface of wings blackish brown, marked with orange patches and with peacock-eye spots; on the fore wings two parallel orange bars cross the cell, and between them and the tip a broad bent whitish band crosses the wing, broadening below and enclosing near the lower outer angle a large peacock-eye with a velvet-black ground; on the outer half of the hind wings are two such spots, the smaller the lower, and between them and the brown margin an orange band. Under surface gray-brown, more or less ferruginous, only the markings of the fore wing repeated, the spots of the hind wing becoming small and inconspicuous ocelli. Expanse more than 2 inches.

Caterpillar.—Head dark glossy brown, sprinkled with yellow tubercles, the summits crowned with an equal spine of moderate height. Body spinous, black-gray, marked with minute, black-edged orange dashes and dots transversely arranged and a pair of maculate pale stripes next the spiracles; spines nearly as long as the segments, all furnished throughout with spinules, not stellate, luteo-fuscous with a metallic lustre. Length 1¼ inches.

Chrysalis.—Brown with dusky shades and more or less mottled and marked with black and cream color, the latter on the abdomen; tubercles and alar ridge blunt and rounded. Length 1 inch or less.

The eggs, which are globose, with ten very thin high vertical ribs and dark green in color, are laid singly on the tips and under side of the leaves of the food-plant and hatch in four days. The caterpillar feeds on Gerardia and a few other Scrophulariaceae, as well as on some other plants, at first upon the under surface leaving only a skeleton, afterwards openly and at all times with no web. The chrysalis hangs from seven to seventeen days, according to the season. The butterfly lives in the open country, has a strong and vigorous flight, and is a southern species, though it is seen occasionally as far north as southern New England and the southern edge of the Great Lakes. In the South there are several broods annually, the butterfly hibernating; in the northern part of its range there may more probably be only two, and it is doubtful whether in the farthest points at which it is found it is indigenous, as all captures have been made late in the season, perhaps the progeny of individuals which have flown far north beyond the natural limits. A single specimen was even taken by Geddes in the Rocky Mountains of Canada, to which it must certainly have flown from a distant point.

11. GENUS VANESSA.

VANESSA CARDUI—THE PAINTED LADY, or THISTLE BUTTERFLY.

(Cynthia cardui, Pyrameis cardui.)

Butterfly.—Upper surface of wings blackish brown, heavily and irregularly marked with orange; apical half of fore wings unequally spotted with white and hind wings with a premarginal series of round black spots. Under surface of fore wings like the upper with exaggerated markings; of hind wings heavily marbled and transversely lined with a mingling of white, olivaceous brown, and gray, the submarginal spots of the upper surface becoming more or less perfect and unequal peacock-eye ocelli, occurring in nearly all the interspaces. Expanse 2⅓-3 inches.

Caterpillar.—Head blackish with pale hairs, not spined on summit. Body spinous, dingy olivaceous yellow, with a more or less inconspicuous delicate tracery of paler color and a mottling of velvety black, varying considerably in relative amount, and with a conspicuous infrastigmatal yellow stripe; spines, including a mediodorsal one on both first and second abdominal segments, yellowish, the spinules of the apical circlet as long as the spine below the circlet; hairs on body much more than half as long as the spines. Length 1¼ inches.

Chrysalis.—Greenish, nacreous, or bluish white, delicately creased with black and banded with light brown or livid, the tubercles often gold-tipped; no distinct supralateral tubercle on eighth abdominal segment, and the wing tubercles blunter than in the other species of the genus. Length somewhat less than 1 inch.

The eggs, which are barrel-shaped, a third higher than broad, with about sixteen thin high vertical ribs and pale green, are laid singly upon the upper surface of the leaves of the food-plant and hatch in from six to eight days. The caterpillar feeds upon almost any kind of thistle, which is its favorite plant, but also upon other Composite plants, especially Anaphalis, and it is also partial to Malvaceae. On hatching the caterpillar leaves its egg-shell uneaten, and after a meal or two on the parenchyma of the upper side of the leaf passes to the under surface and makes a filmy solitary nest next one of the ribs, into which from time to time, as it needs to enlarge it, it weaves bitten particles of the leaf or leaf-hairs; later it makes a larger nest or tent, often at the summit of the plant, sometimes implicating several of its leaves, or stretching across inequalities of surface in a single leaf beneath which it lives. The chrysalis hangs from eight to fourteen days. The butterfly inhabits open fields and is more nearly cosmopolitan in its distribution than any butterfly known, being found in almost every quarter of the globe except in South America (in the northern parts only of which is it found) and the arctic regions. It is generally regarded as single-brooded throughout the greater part of Europe, but with us, even as far north as New England and Canada, it is certainly double-brooded. It hibernates in the butterfly state (perhaps also some autumn chrysalids pass over the winter) and so appears early in the spring. Eggs are laid late in May and early in June; the caterpillars become fully grown between the middle of June and the end of July, and before the middle of July the first brood of butterflies makes its appearance. Eggs are again laid by the end of this month and during August, and late in August or early in September a second brood of butterflies appears. More than most butterflies this species is subject to extensive fluctuations in numbers, and in Europe at least has been known to migrate in vast flocks.

VANÉSSA HÚNTERA—THE PAINTED BEAUTY.

(Cynthia huntera, Pyrameis huntera, Pyrameis virginiensis, Pyrameis terpsichore.)

Butterfly.—Upper surface of wings much as in *V. cardui*, excepting that the largest pale spot in the apical half of the fore wings is white in the male but orange in the female, and that the premarginal series of spots on the hind wing becomes a more or less continuous band with the blue pupil of an ocellus in two of the interspaces. The under surface of the hind wings is smoky brown, with a conspicuous tracery of whitish cross lines on the basal half, and a broad, irregular, mesial white band, beyond which are two moderately large, exquisitely formed, round peacock-eye spots. Expanse 2-2½ inches.

Caterpillar.—Head black, without spines on summits. Body spinous, velvety black, with delicate, transverse, yellowish lines next the incisures, and at the front base of the supralateral spines, from the second abdominal segment backward, a conspicuous, round, silvery-white spot; spines, including a mediodorsal one on both first and second abdominal segments, black; hairs short. Length 1¼ inches.

Chrysalis.—Dull grayish white marked with brown or olivaceous, sometimes golden green marked with purple, the darker markings in part forming an irregular broad band along the sides from one end of the body to the other; tubercles orange-tipped, the supralateral series, including one on the eighth abdominal segment, bluntly conical. Length ¾ inch.

The eggs, which are barrel-shaped, slightly higher than broad, with thirteen to sixteen thin high vertical ribs and yellowish green, are laid singly on the upper surface of the leaves of the food-plant and crowded down between the hairs which cover it; their period has never been recorded. The caterpillars feed almost exclusively on Gnaphalicae, a group of Composite plants nearly allied to the thistles, and particularly on "everlasting," Gnaphalium, but they have also been found on a number of other plants, including thistles. On emerging from the egg, they burrow beneath the silken hairs of the food-plant, bite them off and, mingling them with much silk, form at once a dense white mat; beneath this they devour the parenchyma and then enlarge the nest, never leaving it for food but enclosing larger and larger areas, until finally many leaves are drawn together, the bitten-off inflorescence of the Gnaphalium interwoven with the web, and a nest formed as large as a pigeon's egg; only in the last few days of their life do they leave the nest and devour the entire leaf. The chrysalis, sometimes formed within the final nest, hangs from ten to twelve days. The butterfly is a vigorous flyer and is found in open fields. It is double-brooded in the North, hibernating as a butterfly and also to some extent as a chrysalis. The hibernating butterflies leave their winter quarters about the middle of May and the chrysalids give forth their contents a few weeks later; eggs are laid early in June, and from the middle of July to the end of the first week in August the butterflies of the first brood (proper) of the season make their appearance. Eggs are again laid in August, and the second brood of butterflies flies from the middle of September to the end of the season. As the butterfly is long-lived, individuals may be seen on the wing throughout the entire season from the middle of May to the end of October. In the South the number of broods is certainly greater, and the winter is passed in the butterfly state, if not also in the chrysalis.

VANÉSSA ATALÁNTA—THE RED ADMIRAL.

(Cynthia atalanta, Pyrameis atalanta.)

Butterfly.—Upper surface of wings purplish black, the fore wings with white markings at the apex as in other species of Vanessa, but also with a conspicuous, oblique, curved belt of bright orange across the middle of the wing; hind wings margined with the same. Under surface of hind wings greatly varied with marbling and transverse wavy lineation of pale brown olivaceous gray and black markings of intricate pattern, including a triangular gray patch on the middle of the costal border and a dusting of metallic green on a submarginal series of obscure dark ocelli. Expanse 2¼ inches.

Caterpillar.—Very similar to that of *V. cardui*, including mediodorsal spines on first and second abdominal segments, but perhaps even more variable in coloring; usually, however, more or less of a saffron tint, the distinct light lateral band more commonly macular than in *V. cardui*, the hairs notably shorter, being less than half as long as the spines, and the spinules of the apical circlet not one third as long as the spine below the circlet. Length 1¼ inches.

Chrysalis.—Ashen brown, more or less clouded with blackish fuscous and with a dark stigmatal band, but enlivened by some brilliant more or less golden spots and dotted with black; tubercles brownish yellow except some golden ones in the constricted base of the abdomen, the supralateral series extending upon the eighth abdominal segment and sharply conical. Length more than ¾ inch.

The eggs, which are barrel-shaped with nine thin high vertical ribs and delicate green in color, are laid singly on the upper surface of the food-plant and hatch in five or six days. The caterpillar feeds on Urticaceous plants and almost exclusively on true nettle (Urtica). On quitting the egg the caterpillar partially devours it and then generally makes its way to another leaf—by preference one of the half-opened ones at the summit of the plant—and fastening together different points of the leaf makes a canopy under which it lives, eating only the surface of the leaf beneath the web; later it catches the outer edges of a larger leaf together with silk, and lives in the tube thus formed, devouring the lower edges until it has eaten itself out of house and home; it then forms another nest, first biting the stem partly through so as to cause it to droop. The chrysalis often transforms in one of these bowers after hanging for about ten days. This butterfly, again, is an inhabitant of the open field and is found all over Europe as well as North America. Its life-history is much like that of *V. huntera*, it being double-brooded and hibernating principally as a butterfly, but also as a chrysalis. About the second week in May the butterfly comes out of winter quarters, and by the first week in June the chrysalids begin to disclose their inmates, both sets of butterflies laying eggs at or about the same time, so that caterpillars may be found throughout the whole of June and the first half of July, and butterflies of the new brood emerge from the chrysalis throughout July. Eggs are laid at once, and then a fresh lot of caterpillars may be found directly the old ones have disappeared, or even before that. These develop into butterflies by the very last of August, and continue on the wing until they disappear into their winter hiding-places. This they do among the very last of our hibernating butterflies. Further south there are doubtless a greater number of broods.

12. GENUS AGLAIS.

AGLAIS MILBERTI—AMERICAN TORTOISE-SHELL.

(Vanessa milberti, Nymphalis milberti, Vanessa furcillata.)

Butterfly.—Upper surface of wings blackish brown with two orange fulvous spots in the cell of the fore wings and a very broad premarginal band of the same crossing both wings, on the fore wings divided at its upper extremity; a marginal series of small blue lunules. Under surface slate-brown, the premarginal band gray-brown, crowded with cross-threads of blackish brown, the basal half with distant black cross-threads. Expanse 2 inches.

Caterpillar.—Head black, with white papillae, not spined on summits. Body spinous, the spines shorter than the segments, with a mediodorsal spine on second but not on first abdominal segment; velvety black above, profusely dotted, except on dorsal line, with whitish papillae, giving a snuff-gray appearance, greenish yellow beneath. Length nearly 1 inch.

Chrysalis.—Pale brown, everywhere creased and flecked with dark fuliginous; or pale golden green with indistinct ferruginous creases and then marked with salmon and livid tints; ocellar tubercles pointed, a mediodorsal tubercle on second abdominal segment, the mesothoracic prominence not compressed at tip. Length ¾ inch.

The eggs, which are barrel-shaped, as broad as high, with nine or ten thin and high vertical ribs and pale grass-green, are laid in masses close together in several superposed layers or heaps to the number of several hundred on the under side of leaves of the food-plant near the summit; they hatch in about six days. The caterpillars feed upon nettles and are social in the first half of their life, at once, without devouring the egg-shell, climbing to the summit of the plant, lining it with a web beneath which they swarm; when half grown they disperse and live more openly or in partial shelters, as where three or four may be found together in incompletely closed leaves of nettle, open at tip but closed at base, by which a reversed pocket is formed within which they live when not feeding. The chrysalids usually hang for ten or twelve days. The butterfly has a lively flight, is found by roadsides in Canada and the Northern United States as far south as the latitude of New York City, or higher than that in the Mississippi Valley. It is triple-brooded, hibernating in both the butterfly and the chrysalis state, in the former under piled stones. The wintering butterflies come out while the snow still lies on the ground, and in April the wintering chrysalids give birth to the enclosed butterflies which may be found on the wing through May. Eggs are first laid late in April, and by about the middle of June the butterflies from caterpillars of the same season begin to fly; by the end of July a second, and by the first of September a third brood of butterflies appears, though some of the later chrysalids continue over the winter; even as late as November the butterfly may sometimes be seen on the wing.

13. GENUS EUVANESSA.

EUVANÉSSA ANTÌOPA—THE MOURNING CLOAK.

(Vanessa antiopa.)

Butterfly.—Upper surface of wings rich maroon, deepening into black next the straw yellow, black-dusted, outer margin, and in the black enlivened by small dashes of blue. Under surface nearly uniform black-gray through a mingling of crowded transverse threads of black and blue (as seen under a lens), the broad outer margin ashen white, much flecked with brown. Expanse 3–3¼ inches.

Caterpillar.—Head black, not spined on summits. Body spinous, the spines much longer than the segments, but no mediodorsal spines on either first or second abdominal segments; velvety black, sprinkled with white papillae and with a row of large mediodorsal orange spots; prolegs reddish. Length 2 inches.

Chrysalis.—Dark yellow-brown marked with blackish fuscous, often with a pale bloom and tinged with roseate ; larger tubercles red-tipped ; ocellar tubercles pointed; no mediodorsal tubercle on second abdominal segment. Length 1 inch or more.

The eggs are barrel-shaped, slightly higher than broad, with seven or eight thin high vertical ribs fading next base and are of a pale yellow at first, changing to dark brown and then to inky black; they are laid in a single layer in rings encircling or nearly encircling one of the terminal twigs of the food-plant near its tip and hatch in from nine to sixteen days. The caterpillars feed principally upon willows and elm, but also on poplars and to a less extent on a number of allied plants; they are gregarious throughout life, and in feeding at first range themselves side by side in compact columns; they spin, however but little web and this merely to make a track upon the stems of the food-plant, along which they travel in a procession when moving from place to place. The chrysalis state lasts from eight to sixteen days according to the season, and the butterfly is double-brooded, hibernating in the perfect stage. The butterflies come out the first of the butterfly hibernators— any warm winter day may lure them—and lay eggs early in May, from which a first brood of the season's butterflies springs into being very late in June or early in July; by the middle or last of July eggs are again laid, and the second brood of butterflies is on the wing early in September and remains on the wing until early in November. In the northern part of its range, however, as in the White Mountains of New Hampshire, the butterfly is single-brooded, appearing early in August.

14. Genus Eugonia.

EUGÒNIA J-ÁLBUM—THE COMPTON TORTOISE.

(Vanessa j-album, Grapta j-album, Nymphalis j-album.)

Butterfly.—Upper surface of wings tawny orange, paling into yellow on the outer half of the wings, marked heavily with black especially on the fore wings, where three large black patches depend from the costal margin, while four smaller patches occur in the middle of the lower half of the wing ; a small white transverse bar near apex of fore wings, repeated nearer the base on the hind wing. Under surface brownish cinereous, darker on basal half, everywhere transversely streaked with dark threads or clouded with fuliginous shades ; an L-shaped white spot at apex of cell of hind wings, the lower limb subobsolete. Expanse nearly 3 inches.

Caterpillar.—Head lighter or darker, but dark above and crowned with prominent black spines. Body spinous, variable in color but darker above than below, and more or less green, dotted with white and with longitudinal, light-colored, often whitish, maculate stripes ; the upper spines black with rufous base, the lower lighter colored, those of the thoracic segments with no spinules on the basal half. Length 1½–2 inches.

Chrysalis.— Green of various shades, often covered with a whitish bloom, sometimes clouded with brown, sometimes roseate, the tubercles in the saddle metallic golden, the mesothoracic prominence apically compressed, a mediodorsal tubercle on second abdominal segment, the suprastigmatal tubercle on eighth abdominal segment obsolete. Length 1 inch.

The eggs are doubtless laid in small clusters on the foodplant, but they have never yet been found. The caterpillars feed upon the white birch in company (fifteen have been found together), but no web has been mentioned. The chrysalis hangs for about ten days. The butterfly is a northern species, having in eastern America almost precisely the range of *Aglais milberti*, and is found in forest roads and open woodland. It is probably single-brooded and winters as a butterfly, appearing fresh on the wing at the very end of June and early in July, but becoming much more numerous later in the season and hibernating in October, appearing again in the early spring and laying eggs early in May. A swarm of this butterfly invaded one of the Nantucket light-houses one September night, perhaps in migration.

15. Genus Polygonia.

The butterflies of this genus may be distinguished almost at a glance by their greatly angulated and excised wings. All are tawny-colored above, heavily spotted and, especially the hind wings, broadly bordered with black; the dark markings of the fore wings consist mainly of two stout bars depending from the costal margin and, around the inner bar, of a series of five or six rounded spots arranged in a line bent at right angles, one limb parallel to, the other depending from, the costal margin. The species differ principally in the colorings and markings of the under surface of the hind wings.

POLYGONIA PROGNE—THE GRAY COMMA.

(Vanessa progne, Grapta progne, Grapta c-argenteum.)

Butterfly.—Middle of outer margin of fore wings distinctly crenulate; tail of hind wings not more than twice as long as broad; under surface of same wings gray, traversed by transverse blackish threads, with slight greenish submarginal markings, and a central thin silvery L, the upper limb pointed at tip. Expanse fully 2 inches.

Caterpillar.—Head brown, crowned by long and slender spines having lateral spinules thrown off from the middle, and not so long as the portion of the central spine beyond them;

body spinous, yellowish brown, uniformly variegated above with blackish olivaceous; spines mostly black. Length more than 1 inch.

Chrysalis.—Greatly variegated with buff, olive-green, brown, white and salmon-red; ocellar tubercles equal on basal half, conical beyond, the notch between them broader than deep; largest abdominal tubercles not very much larger than the others. Length nearly 1 inch.

The eggs, which are pale green, barrel-shaped and ribbed, are laid singly on the upper surface of the leaves of the food-plant and hatch in four or five days. The caterpillar feeds openly on species of Ribes (currant, gooseberry, etc.) and probably other Grossulaceae and will eat elm. The chrysalis state varies from ten to sixteen days and has been known to be as short as seven. The butterfly is a northern species, hardly occurring south of lat. 40°, is fond of lanes and the vicinity of barns, and is greatly addicted to the moisture from drying fruit. It is double-brooded, hibernating as a butterfly, coming out in March, laying eggs about the middle of May, and continuing on the wing into June. At the very end of June or early in July the new butterflies begin to appear, lay eggs the same month, and the second brood, which is the more abundant, comes upon the stage in the latter part of August and early in September; very few have not sought their winter quarters by the middle of October.

POLYGONIA FAUNUS—THE GREEN COMMA.

(Grapta faunus, Vanessa faunus, Nymphalis faunus.)

Butterfly.—Middle of outer margin of fore wings conspicuously crenate; tail of hind wings not more than twice as long as broad; under surface of same wings dark gray-brown, much enlivened by green and ashen along the outer third, especially in the male, and with a central, heavy, silvery comma with expanded tips. Expanse fully 2 inches.

Caterpillar.—Head black with a pale W on the front, crowned by not very long black spines. Body spinous, brownish yellow, with a large dorsal white patch on posterior half of body in striking contrast to the rest; spines white. Length 1¼ inches.

Chrysalis.—Pale wood-brown, streaked with dusky green; ocellar tubercles equal on basal half, conical beyond, the notch between them deeper than broad; largest tubercles of abdominal segments not very much larger than the others. Length nearly 1 inch.

The grass-green, barrel-shaped, ribbed eggs are laid singly on the upper surface of the leaves of the food-plant and hatch in one week. The caterpillar feeds principally on willow and black birch, but has also been taken on alder, currant, and wild gooseberry; it does not devour its egg-shell on hatching, but immediately crawls to the under side of the leaf, otherwise living openly and making no sort of nest. The chrysalis state lasts from eight to fifteen days. The butterfly is a northern species, not occurring in the east south of Massachusetts (except along the Appalachians), though in the Mississippi Valley it comes as far south as Iowa and northern Nebraska. It is very active in its movements, partial to roadways, especially through the forest, and although on the wing the entire summer appears to be only single-brooded. It hibernates as a butterfly and lays eggs in the latter half of May and throughout June, and about the middle of July the brood of butterflies of the season appears while some of the hibernators are still on the wing; butterflies continue to emerge from the chrysalis for a month, and it is not until the middle of October that they have all retired to winter quarters.

POLYGÒNIA CÓMMA--THE HOP MERCHANT.

(Vanessa comma, Grapta comma, Vanessa c-album, Grapta dryas, Nymphalis dryas.)

Butterfly.—Middle of the outer margin of fore wings distinctly crenate; tail of hind wings not more than twice as long as broad;

under surface of same wings dark brown on basal half, lighter brown (more or less cinereous in the male) on apical half, considerable variegated (especially in the male) and traversed by short transverse threads of darker brown throughout, with a central heavy silvery comma expanded at the ends. Expanse 2–2¼ inches.

Caterpillar.—Head black, more or less faced with green, crowned by stout and not very long black spines, the spinules of which, emitted from the middle, are about as long as the part of the spine beyond them. Body spinous, varying in different individuals from green to dark brown, in the latter case light below, and transversely and narrowly lined with lighter colors above; spines pellucid. Length 1 inch.

Chrysalis.—Pale wood-brown, tinged and streaked with pale green; ocellar tubercles conical throughout, the largest abdominal tubercles strikingly larger than the others, mesothoracic tubercle triangular on side view. Length nearly 1 inch.

The pale green, barrel-shaped, ribbed eggs are laid singly or more commonly in columns of from two to nine upon the under surface or stems of the leaves of the food-plant and hatch in four or five days. The caterpillars feed on Urticaceous plants, particularly on the hop, to which they are sometimes destructive. The top egg of the column hatches first and the rest in succession down, or rather up, the column; the eggs are not eaten and the caterpillar is strictly solitary, two being rarely found on one leaf; at first it lives openly, but later in life it draws together the edges of the leaf on the under side of which it is living, sufficiently to protect it from sight and the weather, emerging from it at night to feed. The chrysalis generally hangs from seven to eleven days, but late in the season the time is sometimes prolonged to eighteen days. The butterfly is wary and active, inhabits the open country, fields, etc., and is double-brooded. The butterfly hibernates and is on the wing from March to May and sometimes early June, lays eggs on the tender leaves as soon as they burst, and the first fresh butterflies of the season appear at the end of June and fly through August. Eggs are again laid late in July and in August and the butterflies of the second brood appear the last week in August; they have all or almost all gone into winter quarters before October.

There are two very distinct forms of this butterfly, one (dryas) with the upper surface of the hind wings much darker than the other (harrisii); most of the first brood are of the former, most of the second of the latter, but not invariably.

POLYGÒNIA INTERROGATIONIS.—THE VIOLET TIP.

(Vanessa interrogationis, Grapta interrogationis, Grapta fabricii, Grapta umbrosa.)

Butterfly.—Middle of outer margin of fore wings scarcely crenulate; tail of hind wings several times longer than broad; under surface of same wings highly variegated with patches and transverse stripes of various shades of ferruginous brown and ochraceous in the male, nearly uniform reddish brown in the female, in both with a central silver reversed semicolon. Expanse $2\frac{1}{2}$–3 inches.

Caterpillar.—Head lighter or darker brown, crowned by moderately stout spines, the lateral spinules of which are emitted from below the middle. Body spinous, castaneous, uniformly flecked with light dots so distributed as to form longitudinal faintly oblique stripes on each segment; spines luteous or rufous. Length nearly $1\frac{1}{4}$ inches.

Chrysalis.—Various shades of wood-brown tinged with olivaceous, with a fine web of brown in impressed lines, the tubercles of the saddle nacreous; ocellar tubercles conical throughout, the larger abdominal tubercles strikingly larger than the others, mesothoracic tubercle quadrate as seen from the side. Length nearly 1 inch.

The bluish-green, barrel-shaped, ribbed eggs are laid on the under surface of the leaves of the food-plant, either singly or in columns of from three to eight, and hatch in from three to eleven days according to the season. The caterpillars feed upon Urticaceous plants of which hop and elm are the favorites, and also upon linden. They are partially gregarious, several being often found in a loose company; they rarely seek concealment, though they sometimes do so after the manner of *P. comma*. The chrysalis hangs from seven to twenty-six days according to the season and locality. The butterfly is a southern species rarely found north of the Canadian border. In the northern part of its range it is double-brooded, but at least triple-brooded in the Southern States, probably everywhere hibernating as a butterfly; in the region with which we are concerned it leaves its winter quarters early in May and flies until the early part or middle of June, laying eggs late in May and early in June. The first brood of the season's butterflies appears early in July or the last days of June and continues flying until the middle of August; the second brood appears toward the last of August and continues to emerge from the chrysalis even into October.

This butterfly is dimorphic in much the same way as *P. comma*, one form (umbrosa) having the upper surface of the hind wings much darker than the other (fabricii), but differing also in the form of the wings; as in *P. comma* the butterflies of the first brood are mostly of the dark type, but those of the second invariably, or with very rare exceptions, of the lighter type.

Other species of this genus occurring in our district are *P. gracilis*, at the White Mountains of New Hampshire and northwestward; and *P. satyrus*, a Pacific coast species occasionally found in southern Canada.

TRIBE SOVEREIGNS.

16. Genus Basilarchia.

BASILÁRCHIA ÁRTHEMIS—THE BANDED PURPLE.

(Limenitis arthemis, Nymphalis arthemis, Nymphalis lamina.)

Butterfly.—Upper surface of wings velvety chocolate-black, with a broad white bow crossing both wings just beyond the middle. Under surface very dark brown, with a similar bow, a few black-bordered orange spots at the base, and a premarginal series of plain orange spots, besides a double series of crenulate blue lines, next outer margin. Expanse nearly 3 inches.

Caterpillar.—Head dark drab, tuberculate, the summits crowned with a large tubercle, rounded at tip but with raised points; the principal tubercle behind it tumid, but little higher than broad. Body naked, humped, and irregularly tuberculate, of various shades of green, especially olive, with a dorsal patch of pale buff; a pair of long, clubbed, prickly tubercles on second thoracic segment; not more than about twenty minute smooth warts on any one segment above the spiracles. Length nearly 1¼ inches.

Chrysalis.—Varying from creamy white to silvery gray, the wings margined with greenish brown, the body grotesquely streaked; basal wing-tubercle produced to a minute, backward-directed point; tail-piece, seen from above, less than twice as long as its width at apex. Length nearly 1 inch.

The eggs, which are globular, pitted, studded with short filaments, and grayish green, are laid singly on the upper surface of the extreme tips of the pointed leaves of the food-plant, leaves on young plants, only a few feet above the ground, being usually selected; they hatch in from seven to nine days. The caterpillars usually feed upon black and yellow birch, preferably the former, willow and poplar, but have also been found on shadbush and some other plants. As soon as it has hatched the young caterpillar devours its egg, and then begins to feed upon the leaf upon which it was born, beginning at the extreme tip, but always leaving the midrib untouched as it proceeds toward the base; when resting after a meal, it always takes its station on the stripped midrib, to which it fastens with much silk minute bits of leaf to strengthen it; and like all the other species of the genus it makes while young a loose ball of the size of a small pea out of bitten scraps of leaf held together by a few strands of silk and hangs it by a thread or two to the stripped midrib, so that it is moved by every breath of wind—a device, perhaps, to distract from itself the attention of an enemy; for, by constant removals, it is always kept close to the eaten edge of the leaf, while its own perch is as far out on the stripped midrib as it can find a good footing. After the second moult it pays no further attention to this packet, and retires for its siesta to the leaf-stalk or neighboring twig, but it does not quit its feeding spot until the leaf, always excepting the midrib, is almost or quite devoured, when it passes to a neighboring leaf. The chrysalis state lasts from nine to fourteen, usually ten to twelve, days. The butterfly, one of our most striking species, is a northern form, hardly occurring, except in elevated regions, south of New Hampshire, and frequents shaded roads, particularly in the forest. It is perhaps as a rule single-brooded, though a second brood, feeble in numbers, is known to occur; the first brood appears in the latter half of June and remains upon the wing until early in August; the second brood, when it appears, comes very late in August and early in September. The insect hibernates as a half-grown caterpillar, and to do this constructs, like all the species of the genus, a singular hibernaculum: selecting a growing leaf of its food-plant, it eats away the apical third or fourth, excepting the midrib and a narrow flange on each side of it; or it uses the leaf it has been eating, already trimmed in this fashion; it then draws together, above, the outer edges of the uneaten portion to construct a tube, which it lines very heavily with brown silk, within and without; further than this, it binds the leaf-stalk to the stem with repeated windings of silk to prevent its falling to the ground in the winter; by means of the ledge formed by the projecting midrib, it then enters its tube head foremost and completely fills it, so that the opening is just closed by the roughened end of the body. In the spring it quits its winter home as soon as the first tender leaves have appeared.

A form called proserpina, a hybrid between this species

and the next, but more nearly resembling the latter with more or less distinct traces of the white bow peculiar to the former, is found at places along the southern limit of *B. arthemis;* by some it is regarded as a dimorphic form of the present species.

BASILÁRCHIA ASTÝANAX—THE RED-SPOTTED PURPLE.

(Limenitis astyanax, Nymphalis ephestion, Nymphalis ursula, Limenitis ursula.)

Butterfly.—Upper surface of wings blackish, the outer third of the hind wings with three series of pale blue or green spots, the inner of variable width and sometimes suffusing nearly the whole wing, at least in some lights. Under surface brown, with a double submarginal series of blue lunulate lines, a submarginal series of orange spots in a black setting, and a few black-edged orange spots at the base. Expanse 3-4 inches.

Caterpillar.—Head brownish red, tuberculate, the summits crowned with a large nearly spherical tubercle with small projections. Body naked, humped and irregularly tuberculate, strangely streaked, blotched and mottled with brown, olivaceous, and creamy tints; a pair of long, clubbed, and prickly blackish tubercles on second thoracic segment; considerably more than twenty minute smooth warts on most segments above the spiracles. Length 1¼ inches.

Chrysalis.—Grotesquely variegated with patches and streaks of pale salmon, dark olivaceous, inky plumbeous, and yellow-brown, the lighter tints prevailing; basal wing-tubercle rounded or partially suppressed; tail-piece, seen from above, less than twice as long as its width at apex. Length nearly 1 inch.

The eggs, which are globular, pitted, briefly filamentous, and bright yellowish green, are laid as in the last species, but their duration has not been definitely ascertained. The caterpillar is polyphagous, but seems to prefer Rosaceous plants, especially Prunus, Crataegus, and Pyrus; its habits are precisely those of the preceding species in every particular mentioned above. The chrysalis hangs for ten or twelve days. The butterfly is somewhat of a forest species like the last, but not to so considerable a degree, is often found in orchards, and is strangely attracted by a manure-heap. It is a southern species having its northern limits at just about the southern extension of the preceding species. In the north it appears to be partly single-, partly double-brooded, some caterpillars from the first eggs of the season going into their hibernacula when half grown, others continuing to feed, changing to chrysalis and producing a new brood of butterflies late in the season; these lay eggs, the caterpillars from which enter their hibernacula and in the next season develop into butterflies side by side with those from the first brood. The butterflies of the first brood appear in the northern part of their range, i.e., in our district, about the middle of June, continue to emerge from the chrysalis for a month and are still to be seen early in August, about the middle of which month the second, less abundant brood appears and flies through September.

In the South this butterfly is mimicked by the female of *Semnopsyche diana*.

BASILÁRCHIA ARCHÍPPUS—THE VICEROY.

(Limenitis archippus, Limenitis misippus, Limenitis disippus)

Butterfly.—Wings orange with heavy black veins, a broad black outer border enclosing a row of white spots (beneath, a double series of white lunules), a triangular black spot enclosing two white spots and ending in a streak across the fore wings beyond the middle, and, on the hind wings, a heavy, curved, black, extramesial line. Expanse 3-3¼ inches.

Caterpillar.—Head reddish brown, tuberculate, the summits crowned with a large tubercle heavily denticulate at tip, the principal tubercle behind it denticle-shaped, many times higher than broad. Body naked, humped, and irregularly tuberculate, dark olivaceous, often tinged with brownish yellow, and with a cream-colored ragged-edged patch on top of middle abdominal

FAMILY BRUSH-FOOTED BUTTERFLIES. 103

segments; a pair of long, clubbed, and prickly tubercles on second thoracic segment; not more than about twenty minute, smooth warts on any one segment above the spiracles. Length more than 1 inch.

Chrysalis.—Strangely streaked and blotched with blackish green, yellowish brown, pale salmon, and plumbeous, lightest on the abdomen; tail-piece, viewed from above, twice as long as its apical width. Length nearly 1 inch.

The eggs, which are globular, pitted, briefly filamentous, and deep green, are laid as in the other species, but occasionally also on the under surface of the leaf, and hatch in from four to eight days. The caterpillar feeds upon various Salicaceae, particularly willow and poplar; its habits are precisely like those of the other species as recorded above, but it is remarkable that, being everywhere at least double-brooded, the caterpillars of the first brood never form hibernacula, so that we have here an instinct inherited only by alternate generations. The chrysalis hangs from seven to ten days. The butterfly lives in the open country and is widespread; as stated above, it is double-brooded, and probably in the Southern States there is a third brood, which may perhaps sometimes appear as a supplementary feeble brood further north. About the latitude of central New England the first butterflies, from the caterpillars which have hibernated in their first or second, rarely their third, stage, appear the first week in June, continue to emerge throughout this month and begin to lay eggs about a fortnight after they first appear; the second brood appears about the middle of July, while many of the butterflies of the first brood are still on the wing; as butterflies are still to be found laying eggs late in August and even in September, there may possibly be a third brood.

This butterfly has a special interest from its remarkable departure in coloring and pattern from the other species of the genus, thereby mimicking to an extraordinary degree the general appearance of *Anosia plexippus*.

TRIBE EMPERORS.

17. Genus Anaea.

ANÆA ANDRIA—THE GOAT-WEED BUTTERFLY.

(Paphia glycerium, Paphia troglodyta.)

Butterfly.—Fore wings falcate, hind wings tailed. Upper surface either dark orange, margined feebly with brown (male) or paler orange, heavily margined with brown, and with a very irregular, broad, paler band edged with dark brown crossing both wings (female). Under surface nearly uniform dry-leaf brown. Expanse 2½--3 inches.

Caterpillar.—Head gray-green, with minute tubercles which are slightly larger on the summits. Body naked, gray-green, studded with numerous and well-distributed raised paler points. Length 1½ inches.

Chrysalis.—Stout and plump, light green, granulated with white, sometimes speckled with brown, transversely ridged above the wings in the middle of the abdomen. Length nearly ¾ inch.

The eggs, which are nearly spherical, encircled near summit with raised points, and sky-blue when first laid, afterwards turning opaque yellow, are usually laid singly on the under side of the leaf of the food-plant, though often two will be found on a single leaf; they hatch in four to six days. The caterpillar feeds on species of Croton, goat-weed; in its earlier life it devours the tip of the leaf except the midrib, on which it rests as a perch after the manner of Basilarchia, strengthening it by pellets of the leaf attached by silk; after its second moult it lines the upper surface of a leaf with silk, bringing the upper edges together without fastenings, and thus makes a nest like that of Euphoeades, within which it lies concealed, eating the base of the leaf; when this becomes too small it makes a similar nest from another leaf, but goes outside to feed on neighboring leaves, generally toward evening. The chrysalis hangs from seven to twenty days. The butterfly is rapid in flight and shy of approach; it is found in the Mississippi Valley from southern Illinois southward, and west to the Great Plains. The butterfly hibernates early in November, and there are said to be two broods annually, the eggs of the first brood being laid from the middle of May on, of the second apparently in July.

There is said to be "a decided seasonal dimorphism in the two broods of the females."

18. Genus Chlorippe.

CHLORÍPPE CLÝTON—THE TAWNY EMPEROR.

(Apatura clyton, Doxocopa herse, Apatura herse, Apatura proserpina.)

Butterfly.—Upper surface of wings dark tawny marked with blackish brown, the outer half of the fore wings mostly dark, so that the tawny there appears only in two sinuous rows of roundish spots; while the hind wings are wholly tawny except a dark outer margin and a sinuous premarginal row of round black spots. Under surface light brown, with pallid and blackish transverse markings and, on the hind wings only, a sinuous premarginal series of small, nearly round, blue-pupilled ocelli. Expanse 2-3 inches.

Caterpillar.—Head pale green, with two white facial stripes, lateral spines, and the summits crowned by a long spine-like tubercle, having numerous long spinules throughout. Body naked, minutely papillate throughout, striped in green, yellow, and white in continuous and equal bands from head to the forked tail. Length 1½ inches.

Chrysalis.—Pale grass-green, with a yellow stripe marking the dorsal crest which extends the length of the body, and faint oblique stripes on the abdominal segments. Length nearly 1 inch.

The eggs, which are subglobular, with about twenty slight vertical ribs, and yellowish white, are laid on the under side of the leaves of the food-plant in dense patches of from two to five tiers to the number sometimes of five hundred; they hatch in eight or nine days. The caterpillars feed on species of Celtis, the hackberry; they do not devour the egg-shell, and are gregarious in their first three stages, feeding side by side in rows, eating the leaf from the tip backward, but leaving the stouter ribs; they form a pathway of silk wherever they go, but construct no concealment of any kind; after the third moult they disperse and feed singly. The chrysalis state lasts about ten days. The butterfly is a southern species and is therefore found only in the southern part of our district, about as far north as the Ohio River, but occurs in southern Iowa and has been once reported from southern Michigan. It is single-brooded, appearing on the wing in June and July, and the caterpillars hibernate in fallen leaves and crevices of bark at about the time of their third moult.

The species is dimorphic, one form (proserpina) having the upper surface of the hind wings, at least in the female, obscured with brown, while in the other (clyton) it is not so obscured.

CHLORÍPPE CÉLTIS—THE GRAY EMPEROR.

(Apatura celtis, Doxocopa lycaon, Apatura lycaon.)

Butterfly.—Upper surface of wings sordid or gray fulvous, marked much as in the preceding species, but with the lighter spots of the outer half of the fore wings white and therefore very conspicuous, a premarginal ocellus in the lower half of the wing and, on the hind wings, a distinct sinuous black stripe between the dark margin and the row of black spots. On the under surface it differs in a similar way, and also in the larger, more oval, more largely blue-pupilled ocelli of the hind wings, found also to some extent (but usually white-pupilled) on the fore wings. Expanse about 2 inches.

Caterpillar.—Head green, with four pale facial stripes, lateral spines and the summits crowned by a long, apically forked, scarcely spinous tubercle. Body naked, minutely papillate, yellow-green on the back, blue-green on the sides, with faint paler stripes connecting the base of the head tubercles and of the deeply forked caudal spines. Length 1¼ inches.

Chrysalis.—Yellow-green or blue-green, finely specked throughout with pale yellow, with a cream-yellow line along the dorsal crest, which extends the length of the body. Length more than ¾ inch.

The eggs, which are subglobular, with about eighteen slight vertical ribs and pale green, are laid on the under side of the leaf of the food-plant, either singly or in small clusters of a dozen or less, and hatch in three or four days. The caterpillars feed upon Celtis, hackberry, and, lining the upper surface of a leaf so as to cause the sides to curl slightly upward, are partially concealed from view. The chrysalis hangs from seven to ten days. The butterfly is a southern species and extends nearly but not quite so far north as *C. clyton*. It appears to be double-brooded, but some of the caterpillars of the first as well as of the second brood hibernate when half grown and, in the opinion of Edwards, some butterflies also hibernate. The first brood of butterflies of the season appears in June, the second in August; the butterfly life is long, so that some are flying most of the season, while the caterpillars (except those that hibernate) often feed so rapidly that all the earlier stages are passed within a month.

SUBFAMILY MEADOW BROWNS OR SATYRS.

19. GENUS CISSIA.

CÍSSIA EÚRYTUS—THE LITTLE WOOD-SATYR.

(Euptychia eurytus, Megisto eurytus, Hipparchia eurytris, Neonympha eurytris.)

Butterfly.—Upper surface of wings uniform dark brown, with two distant, premarginal, moderately large, circular ocelli, the upper one of hind wings small and inconspicuous, sometimes obsolete. Under surface lighter brown, the ocelli larger, all distinct, more distinctly ringed with yellow, those of the hind wings with satellites; two distant nearly straight brown lines cross the middle of the wings. Expanse 1¾ inches.

Caterpillar.—Head dirty white, heavily mottled with brown, densely papillate, the summits angulate, almost tuberculate. Body naked, but covered with dense pile arising from papillae in transverse series, pale brown with a greenish tinge, with a dark dorsal stripe and obscure brown longitudinal markings; a distinctly constricted neck and short caudal fork. Length fully ¾ inch.

Chrysalis.—Pallid brown, heavily flecked with griseous, the abdomen with a pair of distinct, distant, longitudinal ridges. Length less than ½ inch.

The subglobular, reticulated, very pale green eggs are laid singly on blades of grass, living or dead, and hatch in about thirteen days. The caterpillars feed upon grasses and usually only by night, concealing themselves by day among the roots or on dry sticks on the ground; they are exceedingly sluggish in movement and are lethargic and long-lived, hibernating when more than half grown but not mature. The chrysalis hangs for sixteen days. The butterfly is a southern species, but extends far northward into nearly all the settled parts of Canada except Manitoba, and it has not been reported from Minnesota, though it probably occurs there. It haunts groves and open spots and roads in the forest, is single-brooded, and flies from the last week in May through July, with accessions to the brood certainly through June.

Another species of Cissia, *C. sosybius*, a southern form, occurs as far north as West Virginia.

20. Genus Satyrodes.

SATYRODES EURÝDICE—THE EYED BROWN

(Argus eurydice, Neonympha canthus, Pararge canthus, Hipparchia boisduvalii.)

Butterfly.—Upper surface of wings mouse-brown, beyond the middle paler, especially in the female; a series of four or five small black ocelli distant from the margin. Under surface slaty brown, paler beyond a strongly-waved median brown line, the ocelli repeated, but larger and more complex. Expanse 2 inches or more.

Caterpillar.—Head green, the coronal tubercles very high, conical, red with brown stripes. Body naked, briefly pilose, green, longitudinally striped with darker or lighter green; a distinctly constricted neck and long caudal fork. Length 1¼ inches.

Chrysalis.—Green with buff longitudinal stripes; head acutangulate as seen from sides; abdomen with no longitudinal ridges, the part beyond the wings as long as they are. Length ¾ inch.

The smooth, subglobular, pale green eggs, laid singly, hatch in from seven to nine days. The caterpillars, on leaving them, sometimes devour a part or the whole of the egg-shell and feed on grasses and sedges, having been found on Scirpus and Carex; they feed and mature very slowly, are at first exceedingly sluggish and when not feeding remain on the blade of grass serving as food; but later in life they move about restlessly though slowly and eat with more relish, feeding apparently only by day and mostly in the early morning; they hibernate in the larval condition, nearly grown. The chrysalis hangs for about nine days. The butterfly is found from Iowa to the Atlantic, but does not appear to extend further south than central Ohio and Pennsylvania,* though reaching northward to Hudson Bay. It is found in elevated, moist meadows, and is single-brooded, flying in July and the first half of August.

21. Genus Enodia.

ENÒDIA PORTLÁNDIA—THE PEARLY EYE.

(Satyrus portlandia, Debis portlandia, Hipparchia andromacha.)

Butterfly.—Wings soft brown, slightly paler beyond a median, sinuate (on hind wings doubly arcuate), blackish transverse stripe, beneath with a second nearly straight dark stripe nearer

---

* It has, however, been once taken by Smythe in South Carolina.

the base; a premarginal series of unequal, mostly very large, black ocelli, beneath far more distinctly ocellate than above, and also there encircled with a common pale lilac loop. Expanse 2¼ inches.

Caterpillar.—Head yellowish green, the coronal tubercles moderately high, conical, red-tipped. Body naked, green, sprinkled with very minute white papillae, with a dark green dorsal line and faint side stripes of yellow; a distinctly constricted neck and long caudal fork. Length 1¼ inches.

Chrysalis.—Green, lighter ventrally, the wing ridges creamy; head acutangulate as seen from side; abdomen with no longitudinal ridges, the part beyond the wing-cases much shorter than they are. Length ⅔ inch.

The smooth, subglobular, pure white eggs hatch in from four to six days. The caterpillar feeds on grasses and hibernates when about half grown. The chrysalis hangs for thirteen or fourteen days. The butterfly is a forest species, very gamesome, and has the habit of pitching on tree trunks, head downward. In the North the butterfly is single-brooded, flying from the last of June to the first of August; but in the Southern States it is probably double-brooded, as it appears in West Virginia in the latter half of May, and fresh specimens have been taken in August.

22. GENUS CERCYONIS.

CERCYONIS ALOPE—THE BLUE-EYED GRAYLING.

(Satyrus alope, Hipparchia alope, Minois alope.)

Butterfly.—Wings dark brown, nearly uniform above except for a minute, generally blind, ocellus in the lower median interspace of the hind wings and a pair of distant large black ocelli enclosed in a very broad premarginal yellow band nearly crossing the fore wing. On the under surface the markings of the fore wing are repeated, but the ocellus of the hind wings forms one of a sinuous series of perfect ocelli; while both wings, except the yellow band, are traversed by short transverse dark threads. Expanse 2⅛-2⅜ inches.

Caterpillar.—Head green, papillate, with no summit tubercles. Body naked, finely pilose from minute papillae, green, with a faint slender yellow stripe on the side, the lateral fold also yellow ; no distinctly constricted neck, the tail with a slender but short fork. Length 1⅜ inches.

Chrysalis.—Pea-green, mottled with paler green, the ridges pale straw-yellow, the surface feebly shagreened ; head rectangulate as viewed from the side. Length ⅞ inch.

The eggs, which are short barrel-shaped but tumid, with about twenty-five vertical ribs, and honey-yellow, afterwards pinkish, are laid singly and hatch in from twenty to twenty-seven days. The caterpillars do not devour the egg-shell, but go into hibernation at once upon escape; in the spring they feed upon grasses, but are lethargic and mature slowly, not reaching the chrysalis state until July; this lasts about a fortnight. The butterfly is limited in its northward extension by about the line of the annual isotherm of 45° F., being found in the southern half of New England and westward to Nebraska. It flies in open woods and on the outskirts of shrubbery, is single-brooded, appears about the end of the first week in July and flies into September.

CERCYONIS NÉPHELE—THE DULL-EYED GRAYLING.

(Hipparchia nephele, Erebia nephele, Satyrus nephele, Minois nephele.)

Butterfly.—Differs principally from the preceding species in the total absence of the yellow band of the fore wings, or its substitution by a faint pallid cloud. Expanse 2–2¼ inches.

Caterpillar.—Head emerald-green, papillate, with no summit tubercles. Body naked, finely pilose from minute papillae, dull yellow-green, the sides slightly darker, with a yellow stripe along lateral fold ; no distinctly constricted neck, the tail with a slender but short fork. Length 1¼ inches.

Chrysalis.—Yellow-green with white granulations, the ridges cream-white ; head rectangulate as viewed from the side. Length ⅜ inch.

The eggs, which are like those of *C. alope*, are laid singly and hatch in about twenty-eight days. The caterpillars live on grass and behave precisely as in the other species, and the chrysalis hangs a fortnight. The butterfly flies from Maine to Montana and in Canada, and extends southward so as to overlap a little the northern limits of *C. alope;* it flies in similar places and like it is single-brooded, and in northern New England usually appears about the middle of July and disappears by the end of August.

Along the belt where this species and the preceding overlap, at least in New England, intergrades occur which must probably be looked upon as hybrids.

Cercyonis pegala, by some regarded as a form of *C. alope*, occasionally occurring in New Jersey, is a southern species in which one of the large ocelli of the fore wings is obsolete.

Other genera of this subfamily occurring in our district are: (1) **Neonympha**, of which there are three species: *N. phocion*, a southern species which has occurred, rarely, in New Jersey; *N. cornelius*, also a southern species, taken as far north as West Virginia and southern Illinois; and *N. mitchellii*, known only in southern Michigan and New Jersey. (2) **Coenonympha** with one species, *C. inornata*, a northwestern form which has been taken on Lake Winnipeg and even in Newfoundland. And (3) **Oeneis**, an interesting boreal and alpine genus, of which we have no less than four species: *Oe. calais*, a boreal form found as far south as the southeastern extremity of Hudson Bay and southern Newfoundland; *Oe. macounii*, known only from Nepigon on the north shore of Lake Superior and at the base of the Rocky Mountains in Alberta; *Oe. jutta*, a boreal and circumpolar species which has been taken in some numbers in restricted localities as far south as Ottawa and Quebec in Canada and near Bangor, Maine; and finally *Oe. semidea*, an alpine form found on the barren summits of the White Mountains, N. H., above 5000 feet, and on the highest peaks of the Rocky Mountains of Colorado above 12,000 feet.

Hypatus bachmanii, of the subfamily of Long-Beaks, is a southern species, very erratic in appearance, which has sometimes occurred in considerable numbers in our district, especially in the West, and even so far north as Wisconsin; it has on very rare occasions been taken in New England.

FAMILY GOSSAMER-WINGED BUTTERFLIES.

The subfamily of Erycinids is represented in our district by the genus **Calephelis**, with a single species, *C. borealis*, which has once or twice been taken in New York. All our other members of this family are Lycaenïds.

TRIBE HAIR-STREAKS.

23. Genus Strymon.

STRYMON TITUS—THE CORAL HAIR-STREAK.
(Thecla titus, Thecla mopsus.)

Butterfly.—The hind wings are slightly lobed at the anal angle in the male, rounded in the female. Upper surface of wings uniform blackish brown, the fore wings of the male with a stigma at the end of the cell. Under surface with a sinuous series of very small, pale-edged, black spots across the middle of the outer half of both wings, and, on the hind wings, a submarginal series of larger coral-red spots, bordered within and without with black. Expanse 1¼ inches.

Caterpillar.—Onisciform. Head minute, black. Body naked, with fine pile, dull yellowish green, with a rosy patch on the back of the thoracic and a larger one on that of the hinder abdominal segments. Length ¾ inch.

Chrysalis.—Pale glossy brown, dotted everywhere with dark brown and blackish, the dots forming a faint dorsal stripe on the hinder abdominal segments. Length nearly ½ inch.

This lively butterfly is spread over most of our territory, though rarely found as far north as Canada and never east of western Maine; it is to be found about flowers in open places near thickets. Winter is passed in the egg state, the eggs being deep green, broadly domed, and thickly covered with raised prominences; they are laid singly upon a twig of the food-plant (wild cherry is the only one certainly known, but the caterpillars will eat plum), tucked into some protected spot, and hatch just as the foliage begins to open in the spring. The caterpillar bites a round hole in the top of the egg to escape, does not further disturb it, and at first eats circular holes in the parenchyma of the leaf, then ploughs jagged tracks through it; it will hang by a thread when disturbed, at least when young. It reaches maturity by the last of June or later, the chrysalis state continues for twelve days, and the first butterflies appear about the middle of July; they become abundant by the last of the month, and continue to fly throughout August. There is but a single brood.

24. Genus Incisalia.
INCISALIA NIPHON—THE BANDED ELFIN.
(Thecla niphon.)

Butterfly.—Upper surface of wings dark glossy brown, in the female deeply tinged except at base by ferruginous, the fore wings of the male with an obscure stigma at the end of the cell. Under surface of fore wings yellowish brown with some transverse markings mostly confined to the upper half, according with those of the hind wings, which are cinnamon-brown, crossed before the middle by an exceedingly broad slightly darker band, the borders of which are still darker and very irregular, the outer edged with white; between it and the margin an almost equally irregular series of large ferruginous spots, capped inwardly with blackish. Expanse about 1 inch.

Caterpillar.—Onisciform. Head minute, yellowish brown. Body naked, with fine pile, green, with two distinct whitish-yellow lines along each side. Length fully ⅜ inch.

Chrysalis.—Mingled blackish and yellowish brown, the dark markings of the abdomen extending over the whole surface above the spiracles, the delicate raised reticulation black; a slender dorsal ridge on mesothorax. Length nearly ⅜ inch.

This active butterfly is often seen at a considerable height above the ground, as about the tops of trees, and is to be looked for in open places in the neighborhood of pine woods. In our district it has not been taken west of New York, but it extends north into Canada. Winter is passed in the chrysalis state, and the butterfly, which is single-brooded, appears at the very end of April or early in May and seldom flies beyond this month. The eggs are regularly turban-shaped, rather pale green with white raised reticulation, are laid singly in the latter part of May and hatch in ten days. The caterpillars feed upon pines and one was once found eating into the pod of a garden-pea; they may take a long time to mature, for the chrysalis is sometimes not formed until September.

INCISÀLIA ÌRUS—THE HOARY ELFIN.

(Thecla irus, Thecla arsace, Thecla henrici.)

Butterfly.—Upper surface of wings dark glossy brown, occasionally, especially in female, with slight ferruginous tints, the fore wings of the male with an obscure stigma at the end of the cell. Under surface reddish brown, darkest on basal half of hind wings, the fore wings with slight markings consonant with those of the hind wings, the latter with the basal color outwardly limited by a strongly indented line, beyond which, especially on the inner side, a hoary bloom is conspicuous by a sprinkling of lilac scales; an arcuate series of dusky lunules in middle of outer half. Expanse fully 1 inch.

Caterpillar.—Onisciform. Head minute, yellowish green. Body naked, with fine pile, yellow-green above, red-brown on sides, threaded by a faint green line, green on the lateral fold. Length ½ inch.

Chrysalis.—Black or brown-black with obscure red bands; a narrow black stripe on each side in the middle of the abdomen, not extending to the thorax; a slender dorsal ridge on mesothorax. Length $\frac{8}{10}$ inch.

This butterfly is about the least active of the lively group of Hair-Streaks and is found about shrubbery in roads or open spots. It is a southern form, but occurs as far north as southern Wisconsin in the West and central New York in the East. It hibernates as a chrysalis, and the butterfly, which is single-brooded, appears about the last week in April, the females about a week later than the males, though some do not make their appearance much before June, after the middle of which month they disappear. The eggs are regularly turban-shaped, deep green, with pale-green raised reticulation, and are laid early in June, perhaps earlier, at the base of the flower-stem of the food-plant, and hatch in less than a week. The caterpillar feeds upon the wild plum and possibly other plants, boring into the fruit and inserting its body as far as needed until the entire inside of the fruit is devoured.

INCISALIA AUGUSTUS—THE BROWN ELFIN.

(Thecla augustus.)

Butterfly.—Upper surface of wings dark slate-brown, the fore wings of the male with an obscure stigma at the end of the cell. Under surface of fore wings reddish tawny at base, ochraceous beyond, separated by a nearly straight extramesial brown stripe; of hind wings dark reddish tawny, much infuscated on basal half, which is limited by a deeply indented line; a series of faint dusky dots in middle of outer half. Expanse about 1 inch.

Caterpillar.—Onisciform. Head minute. Body naked, with fine pile, carmine-red. Length $\frac{1}{2}$ inch.

Chrysalis.—Pitchy brown with sparsely-scattered fuscous spots, on the abdomen forming two rows on each side; tracery of raised lines obscure fuscous; a slender dorsal ridge on mesothorax. Length $\frac{2}{3}$ inch.

The butterfly inhabits shrubby rocky heaths, alights by preference on dead vegetation or rocks, a protective resemblance to which will be found in its coloring, and at once on alighting (like many other Hair-Streaks) slides the upraised hind wings repeatedly past each other, while it sidles about in a twitching manner. It is a northern insect found mostly in Canada and extending southward over the whole of New England and along the Appalachian chain, but not known elsewhere in the East. The butterfly is single-brooded and appears from the wintering chrysalis toward the end of April or very early in May, preceding by a few days the last species (where both occur), and flying till the middle of June. Eggs are laid in May or June, but what the caterpillar feeds on is unknown; it probably matures by the middle of July, and the rest of the year is spent in chrysalis.

25. GENUS URANOTES.

URANOTES MÉLINUS—THE GRAY HAIR-STREAK.

(Strymon melinus, Thecla melinus, Thecla hyperici, Thecla favonius, Thecla humuli.)

Butterfly.—Hind wings with a very long thread-like tail and a smaller secondary one. Upper surface of wings bluish black, the hind wings with a large orange lunule seated on a marginal black spot, between which latter and the anal angle is a similar blue-edged black spot. Under surface pearly clay-brown, the hind wings with two orange spots near anal angle, more or less enclosing marginal black spots, separated by blue and interrupting the submarginal series of blackish spots which crosses both wings ; an extramesial series of nearly connected slender black bars edged without with white, within faintly with orange, nearly straight on fore wings, faintly W-shaped on hind wings. Expanse 1¼ inches.

Caterpillar.—Onisciform. Head minute. Body naked, purplish white without markings. Length ⅔ inch or more.

Chrysalis.—Testaceous, discolored and flecked with dark fuscous ; abdomen much wider than thorax, its longest hairs nearly half as long as the segments. Length fully ¼ inch.

This is the only one of our Hair-Streaks which flies almost continuously from May to September; it is found throughout our district, although it has very rarely been taken in any part of Canada; it is to be looked for about shrubbery and vines. The insect is double-brooded and long-lived, which accounts for its continuous presence; it first appears in the early days of May and this brood continues some way into June, while the second brood appears early in July and flies throughout August and sometimes far into September. The eggs are shaped like sea-urchins, and are very delicately reticulate with raised lines and pea-green. The caterpillars feed on the heads of hops and on the pods of beans, Cynoglossum and other plants; they are very active when young and change their form considerably, leech-like, when moving about. It is altogether probable that the insect winters in the chrysalis.

26. Genus Mitura.

MITÙRA DÀMON—THE OLIVE HAIR-STREAK.

(Thecla damon, Thela smilacis, Thecla auburniana.)

Butterfly.—Fore wings of male with a gray stigma at tip of cell; hind wings with a moderately long thread-like tail. Upper surface of wings blackish brown, the larger part of the disk, excepting the veins, dull tawny. Under surface green, the fore wings with a submarginal white stripe edged within with reddish, the hind wings with two basal white bars edged without, and a very tortuous extramesial white stripe edged within, with reddish, besides a slender white margin and a marginal series of powdery spots enlarging toward the anal angle and made up of mingled white, black, and red scales in subocellate form. Expanse fully 1 inch.

Caterpillar.—Onisciform. Head minute, pale green. Body naked, pilose, dark green, with three rows of white or whitish slightly oblique dashes on each side. Length ⅜ inch.

Chrysalis.—Wood-brown, heavily and irregularly marked with blackish fuscous, the abdomen much wider than the thorax, tinged with ferruginous, its longest hairs not more than a third the length of the segments. Length fully ¼ inch.

This is a southern butterfly, flying about as far north as the latitude of 42° and in the West a little further. It seems to occur only in the vicinity of red cedars, on which the caterpillar feeds, and prefers a height of about twenty feet from the ground, near the tops of the cedars, where its active play with its fellows is a very pretty sight. The insect is partly single-, partly double-brooded, and hibernates in the chrysalis state; the earliest butterflies appear about the first of May and continue on the wing throughout June. The eggs, which are turban-shaped with a broad saucer-like depression above, pale bluish green in color and studded with knobs, are laid singly near the tips of the blossoming twigs, tucked into chinks, and hatch in about a week. The caterpillar is of precisely the color of the cedar, feeds on the tips, its head while feeding covered by the segment behind as by a cowl, and takes about five weeks to mature. The caterpillars begin to go into chrysalis toward the end of June; some of these chrysalids hibernate, while others give out the butterfly in about a fortnight, the new brood of butterflies, much less abundant than the first, appearing toward the end of July and continuing through August.

27. Genus Thecla.

THÉCLA LÍPAROPS—THE STRIPED HAIR-STREAK.

(Thecla strigosa.)

Butterfly.—Fore wings of male with a discal stigma; hind wings with a short thread-like tail and the indication of a supplementary one. Upper surface of wings blackish brown. Under surface dark brown, the disk crossed by four subequidistant more or less complete and subcontinuous white threads shifted in position below the median veins, besides the red, blue, and black, white-edged, lunulate marginal markings common to the genus. Expanse 1¼ inches.

Caterpillar.—Onisciform. Head minute, pale brown with a transverse facial black belt. Body naked, pilose, grass-green, very faintly and obliquely striped with greenish yellow. Length nearly ½ inch.

Chrysalis.—Dull yellowish brown, dotted with brownish

fuscous, the reticulation darker; abdomen scarcely wider than the thorax, its hairs half as long as the segments. Length fully ⅛ inch.

This pretty butterfly is widely distributed throughout nearly all our district, failing in the northernmost parts and nowhere very abundant; it has an active nervous flight and is to be looked for in the vicinity of thickets. It is single-brooded, hibernating in the egg state. The eggs are laid on the terminal twigs of the food-plant under the lea of some prominence like a leaf-scar and hatch early in May. The food-plants of the caterpillar are various: thorn, shadbush, and other Rosaceous plants, the common swamp blueberry and doubtless other species of Vaccinium, oaks and willows; Vaccinium and shadbush are probably its favorites. At first the young caterpillar eats little holes through the leaf; afterwards eats holes or bites the edge indifferently, or it may bore into fruit like plums and extract the softer parts; it matures late in June, the chrysalis state lasts from twelve to sixteen days, and the first butterflies appear early in July, sometimes not until the middle of the month, and remain on the wing but a very short time, being rarely seen in August.

THÉCLA CÁLANUS—THE BANDED HAIR-STREAK.

(Thecla falacer, Thecla inorata.)

Butterfly.—Fore wings of male with a discal stigma; hind wings with a short thread-like tail. Upper surface of wings blackish brown. Under surface slate-brown, the disk crossed by four subcontinuous white threads in two distant pairs, the inner pair brief, the outer crossing the wing with tolerable regularity but in a broken fashion, each pair including a darker ground; besides which are the marginal markings peculiar to the genus. Expanse 1⅛ inches.

Caterpillar.—Onisciform. Head minute, very pale green. Body naked, pilose, nearly equal and tapering but little behind, bright grass-green, with lighter and darker green longitudinal lines, or pinkish brown without markings or with heavy dark markings in front and behind. Length ¼ inch.

Chrysalis.—Lighter or darker brown, more or less sprinkled with blackish fuscous dots and blotches and with an obscure dorsal stripe on abdomen; reticulation with larger meshes than in the other species and not elevated at points of intersection; abdomen scarcely wider than the thorax, its hairs not more than one fourth the length of the segments. Length about ⅜ inch.

This butterfly is found about shrubbery in all parts of our district, and is single-brooded, hibernating in all probability in the egg, though eggs have been known to hatch the same season, so that it may also hibernate in an early larval stage. The eggs are pale green, of a turban shape and studded profusely with knobs; they hatch in a few days if in the same season, or mature early in the spring; the caterpillars, which feed on oaks, hickory, and butternut, eat holes in the leaves and mature the last of June and early in July, the chrysalis state continues from fourteen to twenty days, and the butterflies appear at the end of June or early in July, and are to be found through August and occasionally in September. Eggs are known to be laid all through July and early in August.

THÉCLA EDWÁRDSII—EDWARDS'S HAIR-STREAK.

Butterfly.—Fore wings of male with a discal stigma; hind wings with a short thread-like tail. Upper surface of wings very dark brown. Under surface slate-brown, the extremity of the cell marked by a dark bar edged with white, and, besides the marginal markings peculiar to the genus, an extramesial series of transversely oval, dark brown, white-ringed spots. Expanse 1¼ inch.

Caterpillar.—Onisciform. Head minute, black. Body naked, pilose, tapering but little posteriorly, dark brown marked with yellowish brown, with a broad dorsal dark stripe. Length ½ inch.

Chrysalis.—Yellowish brown, streaked and blotched with darker brown, with a dark obscure band on the sides; reticulation with smaller meshes than usual, elevated at points of intersection; abdomen scarcely wider than thorax, its hairs not more than one fourth the length of the segments. Length ⅔ inch.

So far as known, this butterfly inhabits only a narrow strip across the Eastern United States, being rarely found north of lat. 42° or south of 40°; but it is reported in the extreme West beyond our district at widely remote spots, even in the Canadian Rockies. It is an exceedingly lively insect, especially the male, and the story of its life is very similar to that of the last species. It hibernates in the egg state, feeds on oak, biting holes in the leaves, and flies from July to September. As in *T. calanus*, eggs have been known to hatch the same season.

THÉCLA ACÁDICA—THE ACADIAN HAIR-STREAK.

(Thecla californica, Thecla souhegan, Thecla borus, Thecla cygnus.)

Butterfly.—Fore wings of male with a discal stigma; hind wings with a long thread-like tail. Upper surface lustrous dark slate-brown, with an orange lunule on outer margin of hind wings. Under surface pearl-gray with a white-edged narrow bar at end of cell, an extramesial series of white-edged, round, occasionally oval, black spots, and the usual marginal markings of the genus, here more conspicuous, more continuous, and with more orange than usual. Expanse 1¼ inches.

Caterpillar.—Onisciform. Head minute, pale greenish brown. Body naked, pilose, tapering considerably behind, grass-green, with many oblique yellowish stripes on the sides. Length ¾ inch.

Chrysalis.—Dull yellowish brown, spotted with blackish brown, and with a dark dorsal stripe; reticulation with larger meshes than usual, elevated at intersection; abdomen scarcely wider than thorax, its hairs but little more than a fourth the length of the segments. Length ⅔ inch.

The distribution of this butterfly in the East is similar to that of the preceding species except that the belt is removed a little further north, the butterfly being found a short distance only on either side of the Canadian border; it is

FAMILY GOSSAMER-WINGED BUTTERFLIES.

to be found about thickets on the borders of streams where willows, the food-plant of the caterpillar, abound. The caterpillars are very supple in their movements, much like a snail, and eat the willow leaves from the edges inward. The butterfly generally appears just before the middle of July, occasionally earlier, and remains upon the wing during August and possibly later. The eggs then remain unhatched until spring, when the caterpillars attack the tender foliage; they mature at the usual rate, and after from eight to fourteen days in the chrysalis, the butterflies appear.

Thecla ontario is another species of the genus occurring in our district, but is exceedingly rare, and is known chiefly from Ontario and New England; and *T. lorata*, a great rarity known only from Virginia, possibly not distinct from *T. inorata*.

Other genera of Hair-Streaks found in our territory are: **Erora**, with one species, *E. laeta*, a great rarity though found in widely distant places and to be looked for anywhere; **Callicista**, represented by *C. columella*, a species of the Gulf States, once taken at Buffalo, N. Y.; **Calycopis**, with one species, *C. cecrops*, a southern species occurring as far north as Kentucky and West Virginia; **Eupsyche**, one southern species of which, *E. m-album*, has occasionally been taken in New Jersey, Pennsylvania, and Ohio; and **Atlides**, with one species, *A. halesus*, a somewhat common species of the extreme South, which has been taken in Illinois.

TRIBE BLUES.

28. GENUS EVERES.

EVÈRES COMÝNTAS—THE TAILED BLUE.

(Polyommatus comyntas, Argus comyntas, Lycaena comyntas.)

Butterfly.—Hind wings with a short thread-like tail. Upper surface of wings either dark violet (male) or dark brown (female), the hind wings with a marginal series of dark spots, of which the one next the tail is surmounted with orange. Under surface satin-gray, with a very delicate extramesial series of dark brown spots, and with marginal spots much as above. Expanse about 1 inch.

Caterpillar.—Onisciform. Head minute, black. Body naked, pilose, dark green specked with pale points, with a fuscous dorsal stripe, and on either side obscure oblique fuscous markings; last segment broad and flattened. Length nearly ½ inch.

Chrysalis.—Body more than three times as long as broad, pale green, the abdomen brownish yellow, with an interrupted blackish dorsal stripe, and on each side a row of oblique blackish dashes. Length fully ¼ inch.

This butterfly, found everywhere, is a lively insect, often difficult to follow in its motion among the herbage, above which, unless very low, it is seldom seen. Its eggs, which are sea-urchin-shaped, pea-green, and studded with pale prominences, are laid singly, tucked into crevices about the inflorescence of flowers of the Leguminous plants on which the caterpillar feeds—Lespedeza, Desmodium, clover, etc.—and hatch in four days or less; the caterpillar seems to prefer the flower-heads and tender leaves for food and will burrow into the calyx in search of nutriment. The insect is triple-brooded: the first butterflies appear early in May, soon become plenty, and disappear some time in the first half of June; the caterpillars attain their growth rapidly, the chrysalis state is short, and in the first half of July the butterflies of the second brood appear and continue to emerge throughout the month; the same story is again repeated, the chrysalis continuing from nine to eleven days, and the third generation makes its appearance after the middle of August while some worn butterflies of the second brood are still on the wing; the third brood may still be found until after the middle of September. How the winter is passed is not known, but probably as a full-grown caterpillar. Further north it is probable that there are but two broods, as is the case in the White Mountains of New Hampshire.

In southern regions, and as far north as Long Island,

FAMILY GOSSAMER-WINGED BUTTERFLIES. 125

there are two kinds of females, one almost uniformly dark on the upper surface as described above, the other more nearly resembling the male, being blue with broad black margins.

29. GENUS CYANIRIS.
CYANIRIS PSEUDARGIOLUS—THE SPRING AZURE.

(Lycaena pseudargiolus, Cupido pseudargiolus, Polyommatus lucia, Lycaena violacea, Lycaena neglecta.)

Butterfly.—Hind wings with no tails. Upper surface of wings either pale violet with a slight brownish rim or slate-brown (male), or else pallid, more or less tinged with violet, with a very broad brown edging to the fore wings both on costal and outer margins (female). Under surface pale ash-gray with brown markings very variable in extent, especially upon the hind wing, the markings of the disk here varying from a thread terminating the cell and an extramesial series of delicate dots, to a large irregularly-margined blotch covering most of the surface, and only separated from similarly heavy marginal markings by a slender, dentate, extramesial, pallid band. Expanse 1-1¼ inches.

Caterpillar.—Onisciform. Head minute, dark brown. Body naked, pilose, white, with a dusky dorsal line and marked with greenish on the sides ; last segment comparatively slender and but moderately depressed. Length ⅔ inch.

Chrysalis.—Body less than three times as long as broad, light brownish yellow, with a faint dusky dorsal line, and more or less marked minutely with blackish. Length nearly ½ inch.

This highly variable butterfly is found over an immense territory (much more than our district), and the distribution and times of appearance of the different forms which it assumes are mentioned in the Introduction (see p. 18). It occurs in and at the borders of open deciduous woods or by roadsides through them, often settling (with much wavering) in crowds about damp spots. The eggs, which closely resemble those of *Everes comyntas* in color and markings, but are not so flat, are laid singly on the buds or the calyx of the flowers of the plant on which the caterpillar is to feed, tucked in between the flowers well out of sight, and hatch in from four to eight days, according to the season. The plants used as food by the caterpillars are extremely various, those already known belonging to as many as fifteen different families, but their principal food is thought to be Cornus in the early spring, Cimicifuga in June, and Actinomeris later in the season, a plant in, or soon to be in, flower being chosen by the parent; the caterpillars eat buds, flowers, and leaves indiscriminately, but preferably bore into the calyx of flowers and eat out the heart; they are accompanied by ants, which tend them carefully and caress them with their antennae to induce them to emit from their abdominal glands the honeyed secretions thence exuded and which the ants lap up. The butterfly is one of the first to appear fresh from the chrysalis in the spring; the earliest (form lucia) generally appear about the middle of April, and in the first week of May the numbers are materially increased by the advent of the form violacea, and both fly together through this month, further accompanied, after the middle of May, by the third form, neglecta, so that in the last half of this month all may be taken together. In June, lucia is rarely seen and the others disappear one after the other; but in July the second brood proper appears, consisting wholly of neglecta, and continues to emerge from the chrysalis all through this month; it is not so abundant, however, as the preceding, though butterflies may be found even into September. The caterpillars of the second brood when full-fed go into chrysalis, in which state they pass the winter; the summer chrysalids give birth to butterflies generally in ten or eleven days. The above statement is made for southern New England only; there is probably some variation for these dates for places with cooler or warmer climates, for some points regarding which see the Introduction.

FAMILY GOSSAMER-WINGED BUTTERFLIES. 127

Two other genera of Blues also occur in our district, each with two species : **Nomiades**, represented by a boreal species, *N. couperi*, not uncommon about the Gulf of St. Lawrence, and a southern form, *N. lygdamus*, sometimes found in Ohio and even in Michigan and Wisconsin ; and **Rusticus**, likewise represented by a boreal species, *R. scudderii*, taken as far south as Albany, N. Y., and a southern, *R. striatus*, first described from Texas and little known, but said to have been also taken at Racine, Wisconsin.

TRIBE COPPERS.
30. GENUS CHRYSOPHANUS.
CHRYSOPHÀNUS THÒE—THE BRONZE COPPER.

(Polyommatus thoe, Chrysophanus hyllus.)

Butterfly.—Upper surface of wings coppery brown (male) or blackish brown (female), the female with all but the outer border of the fore wings orange fulvous and marked with rows of small black spots which are smaller and obscure in the male; both sexes have an orange band next the outer border of the hind wings. Under surface of fore wings fulvous, of hind wings silvery gray, bordered as above; both wings have a double submarginal series and an extramesial tortuous series of blackish spots, besides a number of others, mostly round, nearer the base. Expanse 1½ inches or more.

Caterpillar.—Onisciform. Head minute, pale. Body bright transparent yellowish green having a velvety appearance, with a dark green dorsal stripe edged with yellow, the whole profusely dotted with minute white mushroom-shaped appendages. Length nearly 1 inch. (From unpublished notes of J. Fletcher.)

Chrysalis.—Light yellowish brown, the abdomen with six longitudinal series of obscure fuscous dots on each side (including those beneath) and a few other dots on the thorax. Length more than ½ inch.

This butterfly, nowhere abundant, is nevertheless found throughout our district except in the eastern half of New England, and eastward; it frequents moist places and flies with less activity than its sprightly allies. It is double-brooded, wintering in the egg state, the butterflies appearing late in June, laying their eggs early in July and continuing through the month. The second brood flies from the middle of August to the middle of September. The pale-green eggs are shaped like a tiny sea-urchin and are laid singly on the seed-pods of the food-plants, Polygonum and Rumex.

31. GENUS EPIDEMIA.

EPIDÈMIA EPIXÁNTHE—THE PURPLE DISK.

(Polyommotus epixanthe, Chrysophanus epixanthe, Lycaena epixanthe.)

Butterfly.—Upper surface of wings dark brown, the male having a burnished chocolate tint with violaceous reflections on the basal half, with three or four blackish dots on the disk. Under surface pale straw-yellow with blackish markings, heavier on the fore than on the hind wings, similar to those of *Chrysophanus thoe*, and on the hind wings a marginal series of slight orange lunules. Expanse fully 1 inch.

Caterpillar and **Chrysalis** unknown.

This is a very local butterfly, found only in peaty meadows, but there often very abundant. It is found all over New England and its borders and near the Canadian boundary westward to the Great Lakes and beyond, but its distribution there is imperfectly known; it is said to have been taken in Kansas. It seems to be single-brooded, appearing at the end of June, continuing to emerge from the chrysalis until beyond the middle of July and flying until the end of the first week in August. The eggs, which are very similar to those of *Chrysophanus thoe*, are laid in July, singly, and apparently do not hatch until the next season. The caterpillar will probably be found to feed upon some dock or knot-weed.

Two other species of Epidemia inhabit our district: *E. dorcas*, found in its northernmost limits, and *E. helloides*, a Pacific coast species reported to be found in Iowa.

32. GENUS HEODES.

HEÒDES HYPOPHLÆAS—THE AMERICAN COPPER.

(Chrysophanus hypophlaeas, Chrysophanus americanus.)

Butterfly.—Upper surface of the fore wings fiery red, the outer border blackish brown; this is reversed on the hind wings, though here the red border is interrupted by dark marginal spots; the fore wings are also furnished with two black bars in the cell and an extramesial series of similar oblique bars. Under surface light brown, tinged on the disk of the fore wings with red and spotted as above; the hind wings are traversed by a submarginal sinuous red stripe, an extramesial sinuous series, and an intramesial straight series of black dots. Expanse 1–1¼ inches.

Caterpillar.—Onisciform. Head minute, yellowish green. Body naked, pilose, grass-green with a faint dusky dorsal line and darker, sometimes roseate, along the middle of the sides. Length nearly ⅜ inch.

Chrysalis.—Light brown or livid, tinged slightly with yellowish green, dotted with blackish, the dots on the abdomen arranged longitudinally in a dorsal series and on either side, above and including the spiracles, five series, sometimes faint. Length nearly ⅝ inch.

This lively and pugnacious butterfly is found everywhere in our district, always in the full sunshine. Even the lovers of nature shut up within the walls of our large cities can enjoy in any public park a sight of these ubiquitous flutterers, can watch them in their hymeneal dance as they toss themselves up and down in contra-unison and then dash away to repeat the sport elsewhere; they are fearless little brilliants and heed not an approaching footstep until just upon them. They are double-brooded in the northern, triple-brooded in the southern, part of our district, changing in New England at about the latitude of Concord, N. H. In the double-brooded district, the first brood usually appears in the first week of June and lasts until the middle of July; the second appears at about the close of the first week of August and flies nearly through September. In the triple-brooded district it first appears about the middle of May and continues nearly to the end of June; the next brood flies from about the end of the first week of July until the middle or latter part of August; the third appears toward the end of August and flies through September. Winter is passed in the chrysalis state, or possibly, in some cases, the full-grown caterpillar may hibernate. The eggs, which are pale green, nearly hemispherical, with very large white-walled cells, are laid singly on the stem or leaf of the sorrel, the food-plant of the caterpillar, and hatch in from six to ten days according to the season. In escaping from the egg, the caterpillar eats only a small hole at the top, and then feeds on the thick parenchyma of the leaf, ploughing its way, first on the under, afterwards indifferently on the upper or the under surface. It goes to the under surface of stones to change to chrysalis, and this state continues, except in winter, from ten to nineteen days according to the season.

33. Genus Feniseca.
FENÍSECA TARQUÍNIUS—THE WANDERER.

(Polyommatus tarquinius, Chrysophanus tarquinius, Polyommatus porsenna, Polyommatus crataegi.)

Butterfly.—Upper surface of wings pale fulvous, broadly and, especially on the fore wings, irregularly marked with dark brown, marginal on the fore wings, basal on the hind wings, varying greatly in the amount of encroachment on the fulvous disk. Under surface pale reddish brown, the fore wings pale on the disk, and both wings, but especially the hind pair, mottled with pretty large, white-edged, dark spots, arranged on the hind wings in transverse series. Expanse nearly 1¼ inches.

Caterpillar.—Head small, pale green. Body largest in the middle and tapering in each direction, naked except for rather short hairs arranged in transverse patches across each segment, and smoky brown marked with smoky stripes. Length nearly ½ inch.

Chrysalis.—Plump with swollen abdomen, which is covered with slight bosses and the hinder extremity flattened and laterally expanded; pallid on the thorax, flecked with brown, dark greenish brown on the abdomen, flecked or blotched with cream yellow. Length ⅓ inch.

This is a southern butterfly, which, however, extends to the northernmost parts of our district in the East, but in the West has not been found nearly so far north. It occurs only in the vicinity of water where alders flourish and is consequently a local insect and flies but short distances. The most remarkable feature in its life-history is the food of the caterpillar, it being the first and almost the only case known among butterflies in any part of the world of a strictly carnivorous habit; its food is confined to plant-lice (aphides) and especially those kinds which exude a fluffy secretion and live in close colonies; into these colonies the caterpillar intrudes, ploughing its way into the mass, and as one after another of the bodies of its victims are sucked dry, their skins are utilized by being involved in the thin loose lining of silken tissue which the caterpillar weaves as it works its way. With a view to this life the butterfly lays its eggs singly upon the twigs of the plant infested by the colonies of plant-lice and in their immediate vicinity or even directly among them. These eggs are of a flattened spheroidal shape with exceedingly delicate reticulation and of a faint green color, nearly pellucid, and hatch in three or four days. The caterpillars attain their growth with unusual rapidity and moult but three times, so that sometimes the chrysalis state is assumed within a fortnight of the laying of the eggs from which the caterpillars are born; the chrysalis, however, hangs an ordinary length of time, from eight to eleven days. In our district there seem to be three broods of this butterfly, which hibernates as a chrysalis, though possibly also as a butterfly; farther south the number of broods is probably greater. With us the first brood flies from the latter part of May to the middle of June; the second brood appears early in July and flies into August; the third from the middle of August until near the end of September.

Another and western genus of Coppers, **Gaeides**, is represented in our district by *G. dione*, which occurs from Missouri to Iowa.

FAMILY TYPICAL BUTTERFLIES.

Subfamily Pierids.

TRIBE RED-HORNS OR YELLOWS.

34. Genus Callidryas.

CALLÍDRYAS EUBÙLE—THE CLOUDLESS SULPHUR.

Butterfly.—Upper surface of wings canary-yellow, the tips of the nervules, especially in the females and on the fore wings, touched with dark brown. Under surface of a similar but less pure color more or less, in the female often very much, marked by scattered flecks of ferruginous in somewhat definite transverse series; at the tip of the cell a more distinct small ferruginous spot, silver-pupilled on the hind wing. Expanse 2½–3 inches.

Caterpillar.—Head pale green. Body naked but sparsely pilose, pale green with a bluish tinge, especially above, and a yellow stigmatal band; each section of the segments with a straight transverse row of small, black, distant papillae. Length 1⅜ inches.

Chrysalis.—Body as a whole distinctly bent in the middle; wing-cases excessively protuberant; frontal horn very long; color usually pale glaucous green with yellow stripes, but sometimes pale yellowish green or roseate, minutely dotted on back with lighter points. Length 1¼ inches.

This is a southern butterfly, very abundant in our Southern States and extending northward into the southern portions of our district, occasionally as far north as southern New York. In the South it sometimes migrates in flocks, apparently always in a southern direction. It seems to be double-brooded, the second brood much more abundant than the first, and as the latter is the only one which has been seen in the North (in August), its occurrence in our district may be entirely due to migration, which its known habits render not improbable; in what stage it hibernates is unknown, but probably as a butterfly, or else as a caterpillar. The eggs, which are yellow, subfusiform, about twice as high as broad and with about seventeen vertical ribs, are laid singly on the more tender leaves of the food-plant, Cassia. The chrysalis hangs ten or twelve days. The male butterfly has an odor like violets.

Two other species of Callidryas occur rarely in the extreme southern limits of our district, in the West: *C. sennae* and *C. philea.*

35. Genus Zerene.

ZERÈNE CAESÒNIA—THE DOG'S HEAD.
(Colias caesonia, Meganostoma caesonia, Zerene cesonia.)

Butterfly.—Upper surface of wings lemon-yellow, the fore wings having the outer border very broadly margined with black, its inner limit so deeply indented, especially in the male, that, with the black dusting of the basal part of the cell and a large round black spot at the tip of the cell, a dog's head is vividly outlined, the round spot forming the eye. Under surface almost uniform yellow, more or less edged and dotted with roseate, the black spot of the fore wings repeated, here with a silvery pupil, and the hind wings with a pair of silver spots enclosed in a roseate or ferruginous nebula. Expanse 2⅛–2¾ inches.

Caterpillar.—Head yellow-green. Body naked but sparsely pilose, yellow-green, usually with narrow transverse bands of yellow or black or both, and studded on each segment with a single transverse series of black or concolorous papillae. Length nearly ¾ inch.

Chrysalis.—Body not bent in the middle, the wing-cases only moderately protuberant, frontal horn short; bluish green with whitish creases and above with two longitudinal rows of black dots, the wings dark green. Length nearly ⅝ inch.

This, a common species in the Southern States and especially in the West, occurs in some abundance in the southernmost parts of our district, and has been found as far north as Pennsylvania, southern Ontario, Wisconsin, and Kansas. The eggs, which are thick fusiform, with about eighteen low vertical ribs and yellow-green in color, are laid singly on the under side of the tender end-leaflets of Amorpha and hatch in about four days. The chrysalis hangs from seven to thirteen days. The butterfly is apparently at least double-brooded and shows some indications of seasonal dimorphism, the later brood or broods having much more roseate on the under surface than the earliest. It is on the wing during every month from April to November, but much is still to be learned of its exact life-history.

36. Genus Eurymus.

EURYMUS PHILÓDICE—THE CLOUDED SULPHUR.

(Colias philodice, Zerene anthyale.)

Butterfly.—Upper surface of wings yellow, the fore wings with a broad, blackish brown outer margin, incurved at the extremities (and in the female broken by yellow spots), together with a small black spot at the tip of the cell; hind wings with a similar border narrowing at the extremities and in the female much narrower and less pure than in the male, in addition to which is a pale orange circular spot at the tip of the cell. Under surface sulphur-yellow, the spots at the tip of the cells repeated, on the fore wings black with a transverse white dash in the centre, on the hind wings ferruginous with a large silver pupil and sometimes accompanied above by a similar satellite. Expanse about 2 inches.

Caterpillar.—Head grass-green with white dots. Body naked, pilose, grass-green, with a faint darker dorsal line and a pale roseate stigmatal band, usually bordered beneath in the middle of most of the segments with velvety black; whole body covered with raised points. Length more than 1 inch.

Chrysalis.—Body not bent in the middle, the wing-cases but little protuberant; frontal horn short conical, the colors on either side of its lateral ridge similar; color of body grass-green, vermiculate with yellowish white, with a narrow yellowish stigmatal stripe. Length $\frac{3}{4}$ inch.

This is our commonest butterfly, found everywhere in open fields, flying rapidly in a zigzag course but little above the herbage, and delighting to assemble in flocks at the edges of pools of standing water, particularly in roadways. It has three broods each year, and probably hibernates as a nearly full-grown caterpillar. The first brood, which is the least numerous, appears at the end of April unless delayed by inclement weather, the males about ten days before the females; its period of greatest abundance is toward the end of May, and early in June only worn specimens can be found; the second brood appears at the end of June and flies until the third brood appears in the latter half of August, and this last is on the wing until the first severe frosts appear. The eggs are laid singly on the upper side of clover-leaves near the middle, and hatch in four or five days; they are fusiform with about eighteen vertical ribs and numerous cross lines; when laid whitish, then faint yellowish green, they turn to a salmon-color, at first faint, afterwards deep, and just before hatching become of a leaden hue. The escaping caterpillar eats its way out at the side, devours a small additional portion of the shell, and then attacks the leaf, resting always upon the midrib while young, on the stalk when older. The chrysalis hangs from nine to eleven days.

The females are dimorphic, many being of a pallid whitish hue instead of yellow, a distinction rarely found in the first brood. One or two instances have occurred of pallid males.

EURYMUS EURYTHEME—THE ORANGE SULPHUR.

(Colias eurytheme, Colias chrysotheme, Colias keewaydin, Colias amphidusa, etc.)

Butterfly.—Differs principally from the foregoing in having the upper surface of the wings orange instead of yellow, and in being tinged with orange beneath. Expanse nearly $2\frac{1}{4}$ inches.

Caterpillar.—Head grass-green with black dots. Body naked, pilose, grass-green with a faint dorsal line and a white stigmatal stripe, which is tracked through the middle by a discontinuous thread of yellow or red and followed beneath by scattered dusky markings, sometimes collected in the middle of the segments into inky spots; whole body covered with raised points. Length 1¼ inches.

Chrysalis.—Body not bent in the middle, the wing-cases but little protuberant; frontal horn short conical, the colors on either side of its lateral ridge contrasted; color of body pea-green, vermiculate with pallid and having a yellow stigmatal band. Length ⅔ inch.

This is a wide-spread and abundant western and southern species, in our district rarely found east of Ohio (though it has been taken even in Maine), with habits like those of the preceding species, but more active in flight and more often flying high in the air. In our district it is triple-brooded, with seasons much as in the preceding species or perhaps a trifle later, and is said to hibernate both as a caterpillar and as a butterfly. The eggs closely resemble those of *E. philodice* but have less numerous cross lines, and hatch in from four to nine days. The caterpillar feeds on clover, and the chrysalis hangs from nine to fifteen days.

This butterfly is remarkable for the extraordinary variety of forms which it assumes, a brief account of which will be found in the Introduction, page 19.

A third species of the genus, *E. interior*, closely resembling *E. philodice* and sometimes mistaken for it, is found in high northern regions, is abundant on the northern shore of Lake Superior, and has occasionally been taken in northern New England.

37. Genus Xanthidia.

XANTHÍDIA NICÍPPE—THE BLACK-BORDERED YELLOW.

(Terias nicippe, Eurema nicippe.)

Butterfly.—Upper surface of wings bright orange, the fore wings with a little dark mark at tip of cell and the outer border broadly margined with blackish brown, which extends above to the middle of the costal margin; in the male it is narrowest in the middle and bends inward on the inner margin; in the female it is broader and fails to reach the inner margin; hind wings with a similar bordering broadest in the middle and, in the female only, nearly obliterated in the lower half. Under surface bright yellow, the fore wings with an orange tinge, the hind wings with some short transverse streaks of ferruginous, especially in the female, where the centre of the disk is often dingy white. Expanse about 2 inches.

Caterpillar.—Head pea-green dotted with black, the papillae high and numerous. Body naked, pilose, the black papillae not transversely arranged, the color green, darkest above, with a broad yellow stigmatal band, edged slightly below with blue. Length 1 inch.

Chrysalis.—Body not bent in the middle, the wing cases very protuberant; frontal horn rather long conical; color green, the raised corrugations white, more or less sprinkled, especially on the wings, with fuscous. Length ¾ inch.

This southern butterfly occurs in the southern part of our district as far north, though not abundantly, as the southern borders of New York; it is common enough in southern but not in northern Ohio. It is found in open fields and has an active flight. It is apparently double-brooded and lives a long time as a butterfly, flying in the South from the time of its first appearance fresh from the chrysalis about the middle of May until the middle of November, with a notable accession in numbers about the middle of August, marking the apparition of the second brood. In keeping with this longevity, the butterfly hibernates and is seen again in the earliest days of spring. The eggs, which are fusiform, with about thirty vertical ribs and of a yellowish-green color, are laid singly (but often many upon the same branch) on the leaves of Cassia and usually upon the under side; they hatch in two or three days. The caterpillars eat first the extreme leaflets of the Cassia, beginning at the tip of the leaf; the chrysalis hangs from five to eight days. As the larval stages are passed rapidly, at least in midsummer, it is possible that the broods may be much more numerous than stated above; but if so, the striking accession to the numbers in flight in August remains to be explained.

38. Genus Eurema.
EURÈMA LÌSA—THE LITTLE SULPHUR.
(Xanthidia lisa, Terias lisa.)

Butterfly.—Upper surface of wings canary-yellow, the apex and whole outer margin (the latter not quite to the outer angle in the female) broadly bordered with blackish brown on the fore wings; hind wings rather narrowly margined with the same in the male, with a large spot at the upper angle only in the female. Under surface duller yellow, sparsely sprinkled with brownish dots, especially on the hind wings, which are more or less flecked with ferruginous and have also a ferruginous spot in both sexes opposite the blackish spot of the upper surface of the female. Expanse 1¼ inches.

Caterpillar.—Head grass-green, the white papillae moderately high and not numerous. Body naked, pilose, the white papillae not in transverse lines; color grass-green, deepening in color down the sides, with a white stigmatal line. Length more than ⅔ inch.

Chrysalis.—Body not bent in the middle, wing-cases but little protuberant; frontal horn slender, conical; translucent green, sparsely dotted with blackish. Length ⅜ inch.

The distribution of this butterfly is almost precisely that of the preceding species, but it has been found in the East a little farther north than it, having apparently a permanent foothold on the southern shores of New England. Probably

FAMILY TYPICAL BUTTERFLIES. 139

triple-brooded in the South, it seems within our district to be only double-brooded, and probably hibernates everywhere as a butterfly. In the North the first brood of fresh butterflies appears about the middle of June and flies for five or six weeks; the second and much more numerous brood appears early in August, receives accessions throughout the month, and flies through September. The eggs, which are light green, slender fusiform, and with very numerous vertical ribs, are laid singly on the upper side of the midrib between the leaflets of Cassia, species with small and finely-divided leaflets being preferred, and hatch in five or six days. The caterpillar escapes from the side of the egg, and generally devours a considerable part of the rest before touching the leaves, when it crawls to the under surface and remains there, at first eating only holes in the leaf so as to leave the skeleton of the leaf untouched; it rests on the midrib of the leaf or on the stalk, and is then difficult to detect, so closely does its color accord with that of the plant; if much disturbed it will drop from the leaf by a thread. In the autumn the chrysalids hang for a month.

An immense swarm of these delicate butterflies, thousands in number, was once blown like a cloud to Bermuda from the mainland, fully six hundred miles away.

39. Genus Nathalis.

NATHÀLIS IÒLE—THE DAINTY SULPHUR.

(Nathalis irene.)

Butterfly.—Wings pale canary-yellow with dark brown markings, which, on the upper surface of the fore wings, consist of a large apical spot bounded by an oblique line connecting the costal and outer margins near their middle, and a broad bar along the inner margin, not reaching the outer margin; this bar is repeated on the under surface accompanied by a couple of blackish spots above its outer extremity; under surface of the hind wings, except next outer margin, much besprinkled with dusky scales. Expanse 1 inch or more.

Caterpillar.—Head green. Body green with a purple dorsal stripe and on each side a double stigmatal stripe of purple and yellow; a pair of reddish, conical, forward-projecting processes on back of first thoracic segment. Length ½ inch.

Chrysalis.—Body not bent in the middle, wing-cases but little protuberant; front rounded, with no distinct horn; yellow-green, thickly dotted with yellow-white. Length ⅝ inch.

This pretty butterfly is very common indeed in the Southwestern States and occurs in our district in southern Illinois and Missouri. Its transformations have been followed by Mr. W. H. Edwards, through whom the above as yet unpublished details are given. The caterpillar feeds on Tagetes, but its seasons are not yet known except that the butterfly flies at the end of June and in July and again very late in the season; doubtless also at other times.

Two other genera of Red-Horns occur in the district: **Phoebis**, with one species, *P. agarithe*, an extreme southern type said to have been taken in Nebraska; and **Pyrisitia**, also with a single southern species, *P. mexicana*, which has been taken occasionally in the West, as far north as Iowa and Wisconsin, and once even in southern Ontario.

TRIBE ORANGE-TIPS.

40. Genus Anthocharis.

ANTHOCHARIS GENUTIA—THE FALCATE ORANGE-TIP.

Butterfly.—Upper surface of wings dull white, the fore wings with a minute black spot at the tip of the cell, the edge of the falcate portion of the wing brown with white dots, and, in the male, the whole apex orange. Under surface of fore wings like the upper, but with no orange tip in either sex; hind wings flecked with light brown collected into large open blotches. Expanse 1¾ inches.

Caterpillar.—Head pallid with greenish inky blotches, crowned with papillae. Body very slender, naked, pilose, numerously striped with orange, green, dark blue, white, and yellow, but principally bluish green, the broader lighter bands being dorsal and stigmatal; numerous black papillae of two different sizes, the larger arranged in series. Length ⅘ inch.

Chrysalis.—Fusiform, pointed at each end; frontal horn plumbeous, thorax pallid, wing-cases yellowish, abdomen pale yellow, the whole dotted with black. Length ¾ inch.

A southern and eastern butterfly, found also in the eastern half of the southern portion of our district, even into New England; it occurs also in southern Illinois and Ohio. It is found in open woods and flies leisurely in a somewhat zigzag course and rarely alights. It is single-brooded and hibernates as a chrysalis. It appears with the first foliage early in May and flies only through this month or for a few days into June. The eggs are tall sugar-loaf-shaped with about fourteen vertical ribs and of an orange color and hatch in four or more days; they are laid singly on the stems and leaves of Cruciferous plants of a slender habit, Sisymbrium and Arabis, and the caterpillars feed on the flowers and buds, and later on the seed-pods. The change to chrysalis is somewhat curious, as related by W. H. Edwards.

Another of the Orange-tips, **Synchloe olympia**, has been found at distant intervals and in scanty numbers in the western and southern parts of our district—Kansas, Nebraska, Missouri, Illinois, Indiana, and West Virginia.

TRIBE WHITES.

41. GENUS PONTIA.

PÓNTIA PROTÓDICE—THE CHECKERED WHITE.

(Piéris protodice, Pieris vernalis, Pieris occidentalis.)

Butterfly.—Wings white, the fore wings marked above with grayish brown by a broad bar across the end of the cell, an interrupted, transverse, unequal belt across the outer third of the wing (subobsolete in the male) and triangular marginal spots at the nervure tips, especially the upper ones; the hind wings have somewhat similar markings in the female. Under surface with similar but heavier markings, both sexes as in the female, but inclining to yellowish brown. Expanse about 2 inches.

Caterpillar.—Head pale straw-yellow, dotted with dark ferruginous. Body slender, naked, pilose, striped with golden yellow and dark greenish purple, dotted with black papillae, which are broader than high. Length nearly 1 inch.

Chrysalis.—With compressed conical elevations above on middle of thorax and on sides of third abdominal segment, the frontal projection not longer than broad, the wing-cases not protruding beneath ; light bluish gray with yellowish dorsal and side stripes and dotted with black. Length ⅞ inch.

This is a southern and western butterfly, flying in abundance farther north in the West than in the East, where it is rarely found north of Ohio, Pennsylvania, and the southern seashore of New England. It has a rapid flight and is most common about vegetable gardens near cities, the caterpillar being destructive to cabbages; since the introduction of *Pieris rapae* to this country, however, it has been largely superseded in this respect by that pest. It is triple-brooded, each succeeding generation more abundant than the preceding, and hibernates as a chrysalis; the first brood appears in May, the second late in June or early in July, the third the last of August. The eggs, which are very tall and regular in form and vertically marked with about fourteen ribs, are laid singly and hatch in four days. The caterpillars feed upon various Cruciferous plants, and in the case of the cabbage devour only the outer leaves of the head and are thus much less destructive in habit than *Pieris rapae*.

The spring butterflies are more heavily marked than those of the subsequent broods.

FAMILY TYPICAL BUTTERFLIES.

42. GENUS PIERIS.

PÌERIS OLERÀCEA—THE GRAY-VEINED WHITE.

(Pontia oleracea, Pontia casta, Pieris napi, Pieris frigida, Pieris cruciferarum.)

Butterfly.—Wings white without markings, or with the veins more or less broadly mapped beneath with gray, especially on the hind wings and on the tips of the fore wings, and the same regions washed with pale yellow. Expanse about 2 inches.

Caterpillar.—Head green. Body slender, naked, pilose, green, minutely dotted with black, except on a dorsal stripe which is not otherwise distinguished. Length ⅜ inch.

Chrysalis.—With compressed conical elevations above on the middle of the thorax and on sides of second and third abdominal segments, those of the third distinctly flaring, the frontal projection much longer than broad, the wing-cases not protruding beneath; color green, the elevated portions infuscated. Length fully ¾ inch.

This northern species occurs throughout all but the southern parts of our region, though in scanty numbers except in mountainous districts; it appears, however, to be absent from the prairies west of the Mississippi, and wherever it has come in contact with *P. rapae*, it has become relatively rare; it seems to be more commonly found in open places in the vicinity of woods than about farms (where *P. rapae* is most common) and is in every respect more feral than the introduced pest. It is triple-brooded, wintering in the chrysalis; the first brood appears at the end of April or early in May, according to the season, and flies somewhat into June; the second at the very end of June or early in July and flies nearly to the end of the latter month; the third early in August or occasionally at the end of July and disappears early in September. The eggs, which are Florence-flask-shaped, tapering from the middle upward and with about thirteen vertical ribs, are pale greenish yellow, and are laid singly on the under surface of leaves, often several on a leaf, and hatch in from five to eight days. The caterpillar feeds on various Cruciferous plants, of which turnip appears to be the favorite, and eats to repletion, the skin of the body being tense and glistening after a meal; it feeds only on the under surface, biting holes through the leaves and never attacking them at the edges. The chrysalis, when not hibernating, hangs from seven to eleven days.

The summer broods are almost pure white beneath, while the spring brood is heavily, often (especially in northernmost localities) very heavily, marked.

PÌERIS RÀPAE—THE CABBAGE BUTTERFLY.

Butterfly.—Wings dull white, the hind wings pale lemon-yellow beneath, flecked uniformly with griseous; fore wings with the extreme apex blackish brown above, more broadly washed with yellow beneath ; besides, on both surfaces is a round black spot on the middle of the outer half of the fore wing and beneath it, on the under surface, a small spot on the inner margin, opposite which, on the costal margin of the upper surface of the hind wings, is a short black bar. Expanse about 2 inches.

Caterpillar.—Head green. Body slender, naked, pilose, green, with a yellowish dorsal band and a similar but slender and interrupted stigmatal band. Length nearly $\frac{4}{5}$ inch.

Chrysalis.—With compressed conical elevations above on the middle of the thorax and the sides of the second and third abdominal segments, the latter not flaring, the frontal projection much larger than broad, the wing-cases not protruding beneath ; color green, the elevated portions infuscated at tip. Length nearly $\frac{4}{5}$ inch.

This butterfly was introduced into this country from Europe at Quebec about 1860, and again at New York in 1868, and has thence spread over our entire region and far beyond, largely displacing our native butterflies, *Pontia protodice* and *Pieris oleracea*, apparently from the earlier appearance of some of the broods and its extreme fecundity; there is no cultivated spot where it cannot be found, and it especially abounds about vegetable gardens both in country and city. It is triple-brooded and hibernates as a chrysalis; the earlier broods appear in any locality where P. oleracea also occurs, about a week or a little less before that species, but the last brood is apparently contemporaneous. The eggs, which are Florence-flask-shaped, taper only on the upper third, have about twelve vertical ribs, are pale yellow and are laid erect in large numbers, but not in close proximity (except accidentally), on the under surface of leaves; they hatch in about a week. The caterpillar feeds on a great variety of Cruciferous plants, especially on cabbage (to which it is very destructive, often totally ruining a crop), but also on some other plants and especially mignonette; in cabbages it bores into the heart and fills the passages with its excrement. When not hibernating, the chrysalis state lasts ten or twelve days. The male butterfly has a very faint but agreeable odor.

Seasonal dimorphism is shown in the heavier markings of the first brood of the season; and a variety sometimes occurs (and was especially prevalent when it was first introduced) of a pale canary-yellow throughout.

Subfamily Swallow-tails.

43. Genus Laertias.

LAÉRTIAS PHILÈNOR—THE BLUE SWALLOW-TAIL.

(Papilio philenor.)

Butterfly.—Upper surface of wings blackish brown with yellowish lunules in the fringe and a submarginal series of pearl-gray spots. Under surface of fore wings nearly the same with larger markings; of hind wings slate-brown at base, beyond varying from metallic green to blue with seven large premarginal rounded orange spots; broadly bordered with black and tipped above with white. Expanse about 4¼ inches.

Caterpillar.—Head black. Body naked, nearly cylindrical, nearly black, with two series of small orange spots on each side, and at both ends of the body, on either side, a series of long black fleshy filaments, those of the first thoracic segment longest. Length fully 2 inches.

Chrysalis.—Body greatly expanded laterally at the third abdominal segment, where the wings form a sharp ridge; a similar compressed ridge on each side of the back of the middle abdominal segments; of a dead-leaf color, more or less infuscated on the elevations. Length more than 1 inch.

A southern butterfly, found over the southern half of our district almost or quite to the southern extremities of the Great Lakes, fond of the blossoms of trees and the damp spots in roads, flying low and rather slowly. It is double-brooded and appears to hibernate as a butterfly, the fresh butterflies appearing in July and September. The eggs, which are subspherical and covered with a gummy red substance, are laid in small clusters, generally of two or three rows of three or four each, but sometimes as many as thirty or more, on the upper side of leaves or more generally on the smaller stems or tendrils of the food-plant, and hatch in from seven to nine days. The caterpillars feed mostly on Dutchman's pipe (Aristolochia), but sometimes on other members of the family, like Asarum, or even on Polygonum and Ipomoea, neighboring plants; they do not devour their egg-shells and at first feed side by side in close company, lying at right angles to the edge of the leaf, heads out; afterwards they are semigregarious, living near together but apart and without concealment. The odor from the scent-organs behind the head is much slighter and less disagreeable than with our other swallow-tails. The chrysalis state lasts three or four weeks.

44. Genus Iphiclides.

IPHICLIDES AJAX—THE ZEBRA SWALLOW-TAIL.

(Papilio ajax, Papilio marcellus, Papilio telamonides.)

Butterfly.—Wings black, transversely marked with broad and narrow whitish stripes, partly common to both wings, narrowing from above downward, and with a large blood-red spot on inner margin of hind wings before the anal angle, generally accompanied within by another. Besides these markings the under surfaces of the hind wings show adjacent median red and white stripes across the wing, and both surfaces blue submarginal lunules in the interspaces below the long tails. Abdomen with yellow sides. Expanse 3-3¾ inches.

Caterpillar.—Head green. Body naked, largest at the third thoracic segment,. pea-green, with transverse markings, consisting of black dots and lines and slender lemon-yellow stripes, besides a conspicuous broad velvety black stripe on the third thoracic segment, edged with lemon-yellow. Length more than 2 inches.

Chrysalis.—Compact, with relatively low prominences except the triquetral elevation on dorsum of thorax; a slender median carina on thorax and a similar pair on upper side of abdomen; dead-leaf brown, or bright green with slight infuscated markings. Length nearly 1 inch.

This butterfly, a southern form, is confined to the eastern half of the continent and is found only in the southern part of our district with about the same limits as the preceding species; it flies low and rapidly among thickets. The insect winters as a chrysalis and has several broods a season; the first brood is dimorphic, one form, marcellus, appearing with the peach-blossoms; the other, telamonides, some weeks thereafter; the second and later broods, ajax proper, also differ from either of the preceding; marcellus disappears about the first of June, telamonides during the same month, while the earliest ajax appear by the time that marcellus has gone, flying with telamonides; thereafter the broods seem to overlap so that they are not easy to distinguish. The eggs, which are oblate spheroidal in shape and pea-green in color, afterward turning black, are laid singly, usually on the upper surface of a leaf and in from four to eight days according to the time of year. The young caterpillar usually devours most of its egg-shell before feeding on the papaw (Asimina) on which it is to live, which it does without concealment. The chrysalis hangs from ten to fourteen days, when it hatches the same season, but an increasing number of each brood do not disclose their inmates at all until the next season; whether this has any definite relation to the dimorphism of the first brood is not yet known. The odor from the scent-organs of the caterpillar is particularly nauseating.

45. Genus Jasoniades.

JASONIADES GLAUCUS—THE TIGER SWALLOW-TAIL.

(Papilio glaucus, Papilio turnus, Jasoniades turnus.)

Butterfly.—Wings bright straw-yellow (paler beneath) with a very broad black outer margin in which are yellow lunules and on the fore wings four black bars descending from the costal margin, the innermost of which, tapering throughout, nearly crosses also the hind wings; besides there is an orange lunule next the anal angle of the hind wings and much dusting with metallic blue, particularly on the under surface on the inner portion of the black border of the same. Abdomen with yellow sides. Expanse 3¾-4½ inches.

Caterpillar.—Head ferruginous. Body naked, largest at the division between thoracic and abdominal segments, deep green, paler below, with a black transverse stripe above at front edge of second abdominal segment, bordered in front by yellow; upper sides of third thoracic segment with a small black-edged greenish yellow spot having a black-edged turquoise pupil. Length 2 inches.

Chrysalis.—Roughened and straight, the wing-cases not prominent beneath, all the higher projections anterior and directed more or less forward; griseous with a yellow olivaceous tinge, often with greenish patches in front and specked and lined with blackish. Length 1¼ inches.

Found everywhere in our district and far beyond it, often swarming in abundance particularly in hilly regions and especially in narrow wooded valleys, often also assembling in vast numbers about damp spots or ordure or decaying animal substances. It winters as a chrysalis and is double-brooded, the first butterflies of the season appearing about the last of May and flying into July, often until the middle of the month, when the second brood, which is less abundant than the first, makes its appearance. The eggs, which are subspherical and leaf-green, are laid singly on the upper surface of leaves and usually hatch in about eight days. The caterpillar feeds on a greater variety of plants than any yet recorded; in all about a dozen families and thirty or more species are already known, among which birch, poplar, ash, and Liriodendron appear to be the favorites; when young it feeds at the edge of the leaf and retires after feeding to the middle of the upper side of the drooping leaf, where it spins a silken carpet whereon to rest head upward; as soon as it moults it chooses a fresh leaf for its residence and spins a new carpet, going to some neighboring leaf to feed; when it grows larger (having moulted three times) it spins a web across a new leaf so tightly as to draw the opposite sides somewhat together and to make of the leaf a sort of trough, the web touching the leaf only at the sides and forming an elastic bed where the caterpillar rests, concealed on a side view. The chrysalis state lasts two or three weeks in the summer.

This butterfly is remarkable for being dimorphic, but with curious restrictions, the dimorphism being limited sexually and geographically; for in the most southern parts of our district and southward there are two forms of female, one resembling the male, as is invariably the case in the north, the other one in which the black has supplanted the yellow to such an extent that the stripes can only be vaguely seen.

46. Genus Euphœades.

EUPHŒADES TROILUS—THE GREEN-CLOUDED SWALLOW-TAIL.

(Papilio troilus.)

Butterfly.—Wings blackish brown, the upper surface with a submarginal series of spots, which are round and pale straw-color on the fore wings, larger, semilunate, and pale blue-green on the hind wings, which have, besides, an orange spot next the middle of the costal margin, an orange and green spot next the anal angle, and the middle of the wing dusted with green and metallic blue in varying quantity. On the under surface of the hind wings this last is replaced by an arcuate series of broad orange lunules, edged within with yellow and without with black, and followed outwardly by metallic blue dusting; but the series is interrupted in the middle by one of the lunules and its appurtenances becoming a comet-like mass of green scales. Expanse about 4 inches.

Caterpillar.—Head pale green. Body naked, largest at the third thoracic segment, dark green, paler beneath, the sides of the third thoracic segment with a large, circular, finely black-edged, buff spot, containing above a small turquoise spot and below a larger velvety black spot; first abdominal segment above with a pair of approximated, finely black-edged, large ovoid buff spots having a small turquoise spot within; and the hinder abdominal segments with transverse series of six small, ovoid, black-edged, turquoise spots. Length 1½ inches.

Chrysalis.—Relatively smooth with no striking prominences except the divergent frontal projections and moderate thoracic elevation; a lateral ridge the whole length of the body, the wing-cases protuberant beneath; pale yellowish green above, all prominences and ridges reddish brown, pale green beneath; or griseous with mingled yellowish and brown dottings above. Length 1½ inches.

This though a southern butterfly extends north to about the 43d degree of latitude, though it appears to be limited westward by about the 95th degree of longitude. Its flight is rather swift and unwearied, in long zigzags, usually only just above the low bushes which it frequents. It winters as a chrysalis and is double-brooded; the first butterflies appear in the last part of May and continue far into July; the second brood is on the wing by the middle of August or earlier, but does not become abundant until toward the end of August. The eggs, which are subspherical and pale green, are laid singly on the under surface of leaves and hatch in probably a week's time. The caterpillar feeds upon various Lauraceae and some other plants, but spice-bush and sassafras appear to be the favorites; after eating its egg-shell it bites a channel through one side of the leaf not far from the tip down to the midrib, and folds the end-flap over to form a concealment; it does not fasten the edge itself in any way, but keeps the flap in place by numerous transverse strands of silk upon the fold of the leaf, and does it so neatly that the edge of the flap just touches the opposite side of the leaf; later in life it brings the two edges of an entire leaf together in the same way and lives therein, feeding upon the neighboring leaves. The chrysalis state lasts about a fortnight.

A second species of Euphoeades, *E. palamedes*, equally common at the South, does not extend so far north as *E. troilus*, but has been taken in Virginia and Missouri and probably may occur at any point on the extreme southern border of our district.

47. GENUS HERACLIDES.

HERACLIDES CRESPHONTES—THE ORANGE DOG, OR GIANT SWALLOW-TAIL.

(Papilio cresphontes, Papilio thoas.)

Butterfly.—Upper surface of wings black-brown with two very arcuate series of very heavy yellow discontinuous markings crossing each other, one passing from the tips of the fore wings to the base of the inner margin of the hind pair, the other, more curved, from beyond the middle of the costal margin of the fore wings to the anal angle of the hind pair, just above which is an orange lunule. Under surface mostly clay-yellow, the markings of the upper surface repeated considerably modified, with great extension of yellow, the hind wings with a median lunulate black belt, marked with blue and centrally with orange. Abdomen with yellow sides. Expanse 4-5$\frac{3}{8}$ inches.

Caterpillar.—Head brown. Body naked, much swollen anteriorly, ferruginous brown with a lateral stripe in front, the hinder end of which (including two or three segments and a broad saddle in the middle) is cream yellow, flecked with brownish, as other parts of the back are slenderly streaked with dirty yellow. Length more than 2 inches.

Chrysalis.—Body roughened and a little bent, the wing-cases protuberant beneath, all the larger projections anterior and directed forward; griseous or dead-leaf brown, often tinged with green and more or less marked with dark brown, especially in front, on the wings except apically, and on the sides of the basal segments of the abdomen. Length more than 1$\frac{1}{2}$ inches.

This largest of our butterflies is a tropical species, but it extends far northward and in recent years has invaded our district, where it is now occasionally found in scattered localities in all the southern portions, having even occurred within thirty miles of Montreal. It rests with its wings expanded and a little depressed and has a sailing flight. It hibernates as a chrysalis and in our district is double-brooded, the first brood appearing early in June and the second at the end of July and much later, flying through September. The eggs, which are subspherical and overlaid by a brownish-yellow secretion, are deposited singly on the tips of the budding leaves in spring, on the older leaves and the twigs later in the year, and hatch in ten or twelve days, or sooner according to some. The caterpillar will probably feed upon any plants of the Rue family and is particularly addicted to the orange, which it sometimes defoliates; it has also been found on plants of allied families; it eats leaves and also the tenderer shoots, and when young remains on the under side of the leaves and devours only the tenderer parts between the ribs of older leaves; later it devours the whole leaf, but even when old it excepts the midrib and rests on the twigs and branches. The summer chrysalids ordinarily hang from six to sixteen days.

48. Genus Papilio.

PAPÍLIO POLÝXENES—THE BLACK SWALLOW-TAIL.
(Papilio asterias.)

Butterfly.—Wings black with markings mainly yellow; fore wings with two straight rows of spots parallel to the outer margin, the outer rounded, the inner triangular; upper surface of hind wings with a median row of spots, in the male forming a maculate band, and a submarginal series of lunules, between which, especially in the female, are many congregated blue scales; at the anal angle a black-pupilled orange demi-ocellus; on the under surface of the hind wings the yellow markings become mostly orange and are heavier. Abdomen with two rows of yellow dots on each side. Expanse $3\frac{1}{4}$–$4\frac{1}{4}$ inches.

Caterpillar.—Head green, broadly striped vertically with black. Body naked, nearly cylindrical, pea-green, marked with black in transverse bands on each segment, broadening into rounded spots at regular intervals by enclosing small, round, yellow spots at their anterior margins. Length nearly 2 inches.

Chrysalis.—Roughened, with the front half bent backward by the protrusion beneath of the wing-cases, all the higher projections anterior and directed more or less forward; dirty yellowish brown, more or less marked with griseous and dotted with black or blackish points. Length $1\frac{1}{4}$ inches.

Found everywhere in our district in cultivated fields and hilly pastures, flying rather swiftly near the ground and often half doubling on its course. Winter is passed in the chrysalis state and there are two broods annually, the first making its appearance in the latter half of May, the second about the middle of July, and each brood flying about two months. The eggs, which are subspherical and honey-yellow, afterward changing in parts to reddish brown, are laid singly on the finely-cut leaves of the food-plant and hatch in from five to nine days. The caterpillars feed on any Umbelliferous plants, and seem to be found on carrot and parsley as often as on anything else; they eat voraciously and live fully exposed, and do not, like most of our Swallow-tail caterpillars, devour their cast skins after moulting. The chrysalis state varies in the summer from nine to eighteen days.

A second species of Papilio, *P. brevicauda*, remarkable for its short tails, has been found in Newfoundland and along the shores of the Gulf of St. Lawrence.

FAMILY SKIPPERS.

TRIBE LARGER SKIPPERS.

49. GENUS EPARGYREUS.

EPARGYREUS TÍTYRUS—THE SILVER-SPOTTED HESPERID.

(Eudamus tityrus, Goniloba tityrus, Thymele tityrus.)

Butterfly.—Upper surface of wings dark chocolate-brown, the fore wings with a belt of four large, contiguous, gleaming, amber-yellow spots, with another smaller one outside of them, and three little fenestrate white spots, one below the other next the costal border a little before the tip. Under surface blackish brown, with a faint gray bloom next the outer margins, the markings of the fore wings repeated, and across the middle of the hind wings, but not reaching either border, a very large unequal silvery white patch. Expanse 2–2½ inches.

Caterpillar.—Head ferruginous with a large orange spot at base of mandibles. Body naked, briefly pilose, greenish yellow, marked with transverse lines, blotches, and dots of grassy green, the lines encircling the body above, the blotches abundant at the sides, and the dots at the anterior edge of each segment; first thoracic segment orange-red with brown shield. Length nearly 1¼ inches.

Chrysalis.—Very stout and plump, the abdomen (exclusive of tail-piece) no longer than the rest of the body; prothoracic spiracle with posterior lip flat; tongue-case not extending beyond the wings; dark brown, marked with blackish and testaceous. Length nearly 1 inch.

This butterfly is found throughout all of our district except the northernmost portions and the eastern provinces; it is found about gardens and has a dashing impetuous flight, starting and stopping abruptly, being perhaps our most robust and vigorous butterfly. It winters in the chrysalis and is single-brooded, although there are two broods in the Southern States, and this may be the case in the southern parts of our district. The butterflies make their appearance with us early in June, sometimes late in May, and continue to emerge from the chrysalis throughout June and fly throughout July and often into August. The eggs, which are domed, heavily ribbed and cross-lined, and of a grass-green color, are laid singly, from the middle of June on, upon the upper surface of leaves, and hatch in about four days. The caterpillars feed upon a number of different plants of the Pulse family (proper), and very likely will eat any of them, but they seem to prefer locusts and especially the rose-acacia; during its first two stages the caterpillar makes a nest by nearly cutting a rounded piece out of a leaf, folding it over and binding the edges to the leaf at a few points with silken cords so that it is open at the sides; when larger it connects two leaves or sometimes more in a similar manner, and often changes to chrysalis therein, first making the nest more secure by a silken interior lining; at other times it makes a cocoon of dead leaves or bits of rotten wood entangled with its silk.

50. Genus Thorybes.

THÓRYBES PÝLADES—THE NORTHERN CLOUDY-WING.

(Eudamus pylades.)

Butterfly.—Upper surface of wings dark glistening brown, the fore wings with a few very small, slender, mostly transverse, fenestrate spots, three just beyond the middle in a triangle, and two sets on the costal margin, one at the middle, the other half way from there to the tip. Under surface as above, but with pale clouds next the margin, and the hind wings crossed by a pair of dark-edged, light-brown, narrow, tremulous bands. Expanse 1¼–1¾ inches.

Caterpillar.—Head black. Body naked, briefly pilose, rather dark green, with a slender darker dorsal stripe, a dull salmon lateral stripe and the infrastigmatal fold pale salmon ; first thoracic segment black, edged in front with red or orange and red on the sides below. Length more than 1 inch.

Chrysalis.—Rather slender-bodied, the abdomen (exclusive of tail-piece) shorter than the rest of the body, the prothoracic spiracle with elevated posterior lip, the tongue-case not extending beyond the wings; fusco-luteous speckled profusely with blackish fuscous, becoming blackish transverse broken bands on the abdomen. Length ¾ inch.

This butterfly occurs throughout our district unless we except the eastern provinces, from which it has not yet been recorded; it is found in open fields and meadows and flies with extreme rapidity and uncertain direction, generally two or three feet only above the ground. It winters as a chrysalis and is double-brooded, the first brood appearing the last week in May, becoming abundant in less than a week, and not wholly disappearing until some time, often late, in July; the second brood is much less abundant than the first, appears in August, usually not until the middle of the month and flies till the middle of September or later. The eggs, which are subglobular but with a broad base and with moderately low vertical ribs to the number of fifteen, are very pale green, almost white, and are laid on the under side of leaves, singly, and hatch in from five to eleven, generally about six, days. The caterpillar feeds on almost any Leguminous plant, but appears to prefer clover and bush clover (Lespedeza); on emerging the caterpillar usually devours about half its egg-shell and then travels to another leaf to prepare its nest, which it makes by cutting two parallel channels inwards from the edge of the leaf and folding over and securing by silken strands the flap thus formed; later in life it makes a larger nest from one or more leaves after the habit of Epargyreus; it is very cleanly, always ejecting its excrement outside its nest with a snap which sends it to a distance. The chrysalis state in summer lasts about twenty days.

A southern species of Thorybes, *T. bathyllus*, very close to this but with larger spots, has been occasionally found far north, even as far as Massachusetts and Wisconsin ; and another species, *T. electra*, is known only from Hamilton, Ontario.

51. Genus Thanaos.

THÁNAOS LUCÍLIUS—LUCILIUS'S DUSKY-WING.

(Nisoniades lucilius.)

Butterfly.—Upper surface of wings dark grayish brown, the fore wings with the basal half blackish, a double row of premarginal gray spots and, next the costal margin beyond the blackish base, a large and distinct cinereous patch, followed outwardly by three minute vitreous spots one above the other. Under surface fuliginous brown with pallid spots and dots in submarginal series. Expanse about 1⅜ inches.

Caterpillar.—Head with the summits considerably elevated, black with three reddish spots and streaks on the sides. Body naked, briefly pilose, with a slender pale yellowish lateral line. Length ⅘ inch.

Chrysalis.—Slender, the abdomen (exclusive of tail-piece) longer than the rest of the body, the posterior lip of thoracic spiracle scarcely raised, not flaring, the tongue-case scarcely extending beyond the wings ; pale green. Length somewhat more than ½ inch.

This butterfly probably occurs throughout our district, but it has never been reported from Canada excepting in southern Ontario, nor west of this locality except in distant Dakota; nor in New England, where it is best known, has it been found north of Plymouth, N. H., nor in Maine or beyond that; it occurs in wooded rocky spots and winters as a full-fed caterpillar. It is partly single-, partly double-, and partly triple-brooded, there being annually three apparitions of the butterfly in decreasing numbers: early in May, the middle of July, and the middle of August, some of the caterpillars of each of the first two broods as well as all of the last ceasing to feed after they are full grown and passing the winter in the larval nest, closing it tightly, and only changing to chrysalis very early in the following spring; but the last brood of the season is made up not only by direct descent from the second, but also by a certain proportion of the lethargic caterpillars of the first brood, which, when the regular time for change in the second brood of caterpillars occurs, change *then* to chrysalis, instead of doing so as soon as full fed or of waiting still longer until the succeeding spring. The eggs, which are subspherical with broad base and twelve to fifteen compressed and not very high vertical ribs, are at first whitish green, afterwards salmon-red, and are laid singly on the under surface of the leaves of the food-plant and hatch in about ten days in June. The caterpillar feeds on wild columbine, *Aquilegia canadensis*, and has also been found in the South on *Chenopodium album;* although it does not eat much of its egg-shell, it generally takes the caterpillar about twenty-four hours to eat its way out, and this deliberate manner it retains through life; it makes its first nest much after the manner of Thorybes, and after it has bitten the channels requires three or four hours of continuous work to bring the flap to the proper angle required for the nest; when it leaves a nest to form a larger one it always first bites off the strands which have kept the old flap in place; it goes to another leaf to feed, and when mature makes a nest of a whole leaf or of several leaves. The chrysalis state in summer lasts from eleven to fifteen days.

THÁNAOS PÉRSIUS—PERSIUS'S DUSKY-WING.

(Nisoniades persius.)

Butterfly.—Upper surface of fore wings dark grayish brown, the basal half and a band across the middle of the outer half blackish; between the two, next the costal margin, an indistinct cinereous patch, followed outwardly by a descending row of four or five minute vitreous spots; hind wings chocolate-brown.

Under surface dark fuliginous, with the vitreous spots of the fore wings repeated and a cinereous apex. Expanse about 1⅜ inches.

Caterpillar.—Head with summits rounded and slightly elevated, ferruginous brown with pale vertical streaks, or piceous marked with ferruginous. Body naked, briefly pilose, pale green with pale yellowish lateral lines and sprinkled profusely with white dots. Length more than 1 inch.

Chrysalis.—Dull olivaceous green, much infuscated, abdomen pinkish brown, mottled faintly with pale dots, the rest as in the preceding species. Length somewhat more than ½ inch.

Probably found over the whole of our region, but not yet noticed in the northernmost portions; it occurs mostly in shady roadsides by woods and is strong, rapid, and restless in flight, often flying in little circles as if about to alight and then darting off again. It hibernates as a full-grown caterpillar and changes to chrysalis before vegetation has started. It first appears as a butterfly early in May and continues to emerge from the chrysalis throughout the month, after spending sixteen days or more in chrysalis; by the middle of June it has disappeared. It is possible that there is a second brood, as fresh specimens have been taken in the latter half of July; but if so, it is but a small one and the insect partly single-, partly double-brooded, most of the caterpillars of the first brood remaining unchanged until the succeeding year. The eggs, which are shaped as in the preceding species with from eleven to fourteen vertical ribs, more elevated above than in *T. lucilius*, are yellowish green in color, changing afterwards to blood-red; they are laid singly on the upper surface of tender terminal leaves and hatch in about a week. The caterpillar feeds upon willows and poplars, and on emerging from the egg eats only the crown; it constructs a flap-nest like the last species, the flap being at first folded downward, later ones upward; when very young it eats only the parenchyma of the surface of the leaf near its nest; later little holes through the leaf, giving it a riddled appearance; when half grown it always rests with the two ends of its body bent at right angles.

THÁNAOS JUVENÀLIS—JUVENAL'S DUSKY-WING.

(Nisoniades juvenalis, Nisoniades ennius.)

Butterfly.—Upper surface of fore wings dark grayish brown, paler in the female, much besprinkled with gray scales, with a vitreous spot at tip of cell and a transverse series of similar spots in the middle of the outer half, interrupted beyond the cell, and those beneath duller, all set in a broken obscure blackish band, distinct only at their margins; hind wings cloudy blackish brown, the outer half obscurely marked with slightly paler spots. Under surface dark purplish brown with a grayish tinge, the spots of the upper surface repeated more distinctly, and besides, on the hind wings, a pair of small brown-edged yellowish spots near upper outer angle. Expanse about 1¾ inches.

Caterpillar.—Head with summits rounded and somewhat elevated, varying from greenish fuscous to fawn color, heavily marked on the sides with pale orange. Body naked, briefly pilose, light or dark green, with slender pale-lemon lateral stripes, and dotted profusely with pale yellow. Length 1 inch or a little less.

Chrysalis.—Pale or livid above, the abdomen faintly tinged with salmon above and below, the metathorax slightly infuscated; all the appendages in great part black or blackish fuscous, the disk of the wings dark olivaceo-fuscous, the rest as in the other species. Length more than ½ inch.

Found throughout our district, except in some of the northernmost portions, in open oak thickets flying vigorously. The winter is passed as a full-fed caterpillar and the species is probably both single- and double-brooded, the second brood of butterflies being very much less numerous than the first. The butterflies first appear on the wing at the very beginning of May and fly until the middle of June, being most abundant about the middle of May; the second brood appears after the middle of July and flies through August. The eggs are shaped as in the other species, with about sixteen vertical ribs, highest above, and are pea-green, changing after two days to a salmon-red; they are laid singly on the stems and perhaps also on the leaves of the food-plant and hatch in eight or nine days. The caterpillar feeds principally upon oaks, but also upon some Leguminous plants, and makes a nest like the preceding species, but always, even when young, travels to a distance for its food. When winter approaches, the hibernating caterpillar takes on a vinous tint. In the spring the chrysalis state lasts a full month.

THÁNAOS BRÌZO—THE SLEEPY DUSKY-WING
(Nisoniades brizo.)

Butterfly.—Upper surface of wings very dark grayish brown, the fore wings flecked with white scales especially toward the apex, with no vitreous spots, but crossed by two distant dark bands with jagged black edges, the outer band the more distinct; hind wings with a few small obscure pallid spots on outer half. Under surface dark fuliginous brown, the fore wings gray apically and both with a marginal and premarginal series of small whitish spots. Expanse somewhat more than 1¼ inches.

Caterpillar.—Head dark brown, paler above, with an orange spot at base of mandibles. Body naked, briefly pilose, pale green with an indistinct paler lateral stripe and dotted with darker green. Length more than 1 inch.

Chrysalis.—Green, the appendages infuscated, the rest as in the other species. Length ½ inch.

Occurs in every part of our district in moist shady spots and forest openings, flying swiftly about three feet from the ground with sudden lateral movements. It hibernates as a full-grown caterpillar and is single-brooded, appearing on the wing very early in May, becoming abundant by the tenth of the month and flying until the middle of July. The eggs, shaped as in the other species, have fifteen ribs of uniform height. The caterpillar feeds upon scrub oak and perhaps Galactia, and its habits are in general like those of the other species of the genus. In the spring the chrysalis state lasts but nine days in the Southern States, probably longer in the Northern.

THÁNAOS ICÈLUS—THE DREAMY DUSKY-WING.
(Nisoniades icelus.)

Butterfly.—Upper surface of wings very dark grayish brown, the fore wings heavily flecked with cinereous, especially on the apical half and in a large roundish patch next the costal margin between the two dark bands which traverse the wing and which it shares with *N. brizo*, but the inner of which is usually less distinct than in that species; between the outer band and the margin is a uniform series of small round brown spots; otherwise as in *N. brizo*. Expanse 1¼ inches.

Caterpillar.—Head light reddish brown, with slightly raised summits. Body naked, pilose, pale green dotted with white, giving a gray-green appearance, and with a pallid lateral stripe. Length nearly ¾ inch.

Chrysalis.—Anterior portion of body reddish or yellowish brown, the abdomen pale flesh-color, the rest as in the other species. Length fully ½ inch.

Found everywhere in our district in damp wooded regions, especially among the hills, rarely flying at all in companies. It is single-brooded and hibernates as a full-fed caterpillar, changing to chrysalis in the spring, remaining in that state at least two or three weeks and appearing on the wing about the middle of May; it becomes quickly abundant and flies until and into July. The eggs are very pale green with from ten to fourteen vertical ribs, highest above, and are laid singly upon the upper surface of leaves, those tolerably young but fully expanded being preferred; they hatch in about ten days. The caterpillars feed upon aspen, willow, and witch-hazel, and make nests like the other species, but with the attaching strands of silk unusually long, shortening them when they wish to change their skin within before desertion for another nest; they line this nest within with silk for winter quarters. All the species of Thanaos rest with fully expanded wings.

Other species of Thanaos that occur within our district are *T. horatius*, a southern form which has been found along the Atlantic coast as far north as Massachusetts, but is very rare; *T. terentius*, a much rarer species, of which the same may be said; *T. martialis*, a wide-spread species occurring in at least the southern half of our district from Massachusetts to Kansas, but which seems to be nowhere common except in the Southern States; and *T. ausonius*, which is so far certainly known only from Albany, N. Y.

52. Genus Pholisora.

PHOLISORA CATÚLLUS—THE SOOTY-WING.

(Nisoniades catullus.)

Butterfly.—Wings nearly black, the fore wings with an oblique descending series of three small white spots just before the tip, followed by an arcuate series of five white dots beginning at right angles with the former (frequently obsolete beneath), and a similar white dot in the cell. Expanse 1¼ inches.

Caterpillar.—Head black, summits rounded. Body naked, briefly pilose, dull pale green; thoracic shield velvety black, slender, pallid at the edges; second pair of legs resembling the third pair more than the first. Length ⅝ inch.

Chrysalis.—Body slender, the abdomen (exclusive of tail-piece) longer than the rest of the body, posterior lip of thoracic spiracle elevated, flaring; equal apical portion of tail-piece as seen from above scarcely longer than broad; color yellowish green, with brownish dorsal line, and similar ventral line on abdomen. Length ½ inch.

Found in all our district except perhaps some northernmost portions, from few of which it has been reported, flying in gardens and fields. It hibernates like the species of Thanaos as a full-fed caterpillar and is apparently double-brooded in our district, but triple-brooded in the Southern States; it first appears about the middle of May and again late in July, then flying until September. The eggs are very broad sugar-loaf-shaped, broader than high and with vertical ribs which are very coarse and thick at summits, of a yellow color inclining to carneous, and are laid singly on the upper surface of leaves; they hatch in about five days. The caterpillar feeds principally upon Chenopodiaceae and Aramantaceae, especially Chenopodium and Aramantus; when young, nests are made like those of the young Thanaos; later a whole leaf is used, bent at the midrib and the edges fastened at wide intervals by very short strands of white silk; these nests are entirely closed with silk previous to a moult, and similarly closed and lined when prepared for the winter's sojourn. The chrysalis state lasts seven or eight days.

Another species of this genus, *P. hayhurstii*, found in the Southern States, occurs as far north as Kansas, West Virginia and Maryland.

53. Genus Hesperia.

HESPÈRIA MONTÍVAGA—THE VARIEGATED TESSELLATE.

(Pyrgus montivagus, Hesperia tessellata, Syrichtus communis.)

Butterfly.—Upper surface of wings blackish brown, largely checkered with white spots, prominent among which is a broad median series of squarish spots, longer than broad, a premarginal series of small triangular or squarish spots, followed by a row of dots; and on the fore wings, between the two principal series in the upper half of the wing, two series of elongate white spots. On the under side of both wings these markings are repeated, but on the hind wings, the ground of which is greenish brown, there is also a basal white band. Expanse 1¼ inches.

Caterpillar.—Head piceous, the summits rounded. Body naked, briefly pilose, green with a dark interrupted dorsal line, dark lateral bands, and a pallid band below the spiracles; thoracic shield blackish brown; second pair of legs resembling the first pair rather than the second. Length ⅜ inch.

Chrysalis.—Body slender, the abdomen (exclusive of tail-piece) longer than the rest of the body, posterior lip of thoracic spiracle elevated, flaring; equal apical portion of tail-piece as seen from above twice as long as broad; yellowish white, dotted above with black. Length nearly ⅜ inch.

A southern and western species found in nearly or quite all the western part of our district (but sparingly in the North), and in the East hardly occurring north of southern Ohio and Pennsylvania; in the far West it is perhaps the commonest of butterflies; its flight is very rapid and close to the ground. Its life-history is insufficiently known, but it appears to winter in the chrysalis and to be triple-brooded, the successive broods appearing early in spring, again in June and July, and once more, and more abundantly, in August and September, actually flying continuously from early spring until late autumn. The eggs which are nacreous-white, nearly spherical, with twenty-four prominent vertical ribs, are laid singly upon the upper surface of leaves. The caterpillar feeds upon various mallows: Sida, Malva, Althaea, and Abutilon. In summer the chrysalis state lasts from eight to twelve days.

Another species of Hesperia, *H. centaureae*, a high boreal and circumpolar form, has been taken in one or two instances in the extreme east of our district even as far south as West Virginia.

Other genera of Larger Skippers found in our district are **Eudamus**, with one species, *E. proteus*, a tropical type occasionally found on the Atlantic border as far north as New York; **Achalarus**, represented by *A. lycidas*, a southern form which has been occasionally taken in Wisconsin, Illinois, Michigan, New Jersey, New York, and southern New England; and **Rhabdoides**, with one species, *R. cellus*, again a southern type which is found at least as far north as West Virginia and Kentucky.

TRIBE SMALLER SKIPPERS.
54. Genus Ancyloxipha.

ANCYLÓXIPHA NÙMITOR—THE LEAST SKIPPER.

(Thymelicus numitor, Heteropterus marginatus.)

Butterfly.—Antennal club with no recurved hook at tip Upper surface of wings tawny, very broadly bordered with dark brown, the fore wings so broadly as to be almost wholly brown; male with no discal dash. Under surface golden tawny, all but the broad costal and outer margins of fore wings blackish fuliginous. Expanse about 1 inch.

Caterpillar.—Head blackish brown. Body naked, pale greenish yellow, dotted with fuscous, the thoracic shield brownish fuscous (immature; full-grown caterpillar unknown).

Chrysalis.—Reddish ash color, minutely sprinkled with brown dots, the tongue-case reaching the base of the tail-piece.

Known from all but the northernmost portions of our district, northern New England and the Eastern Provinces; it occurs in the vicinity of running water and in marshy meadows and flies in a languid leisurely manner close to the ground. It is triple-brooded and passes the winter either as a mature caterpillar or as a chrysalis, probably the latter. The butterflies come early in June and disappear before the end of the month; again late in July, disappearing by the middle of August or soon after it; and once more in the last week of August, flying nearly to the end of September. The eggs, which are low hemispherical, smooth and shining yellow, afterward orange-red, are laid singly and hatch in from five to ten days according to the season. The caterpillar feeds upon common grasses, probably in nature upon some semiaquatic species; when first hatched it makes a nest in a blade of grass by pulling the edges partially together with five or ten strong strands of silk, broadest at their bases, and lives behind the strands; later it fills in the interstices with a finer web. The chrysalis state in summer lasts in Georgia about ten days.

55. Genus Atrytone.

ATRYTONE ZÁBULON—THE MORMON.

(Pamphila zabulon, Hesperia hobomok, Hesperia pocahontas.)

Butterfly.—Upper surface of wings blackish brown, heavily marked centrally with tawny, forming on the hind wings a large, central, more or less angular patch, on the fore wings a number of irregular and very unequal spots in the interspaces ; male with no discal dash. Under surface dark cinnamon-brown, on the outer margin flecked with lilac, and centrally marked heavily with lemon-tawny as above, but the markings on the fore wings are blended with an oblique black line at the end of the cell, and on the hind wings form a definite transverse band abruptly and considerably broadened in the middle. Expanse about 1⅜ inches.

Caterpillar.—Head dark ferruginous, scabrous. Body naked, briefly pilose, yellowish brown, with dark dorsal and lateral lines and dotted with fuscous ; a narrow, interrupted, fuscous thoracic shield, in front of which the segment is greenish. Length ⅘ inch.

Chrysalis.—Uniformly livid, somewhat infuscated on head and thorax, the appendages with a whitish bloom ; tongue-case extending to the eighth abdominal segment. Length nearly ⅘ inch.

This butterfly is found throughout our district, in meadows, flying swiftly and abruptly, close to the ground. It is single-brooded and passes the winter sometimes as a full-grown caterpillar, sometimes as a chrysalis. The butterfly appears the last week in May, becomes abundant early in June, and disappears before the end of that month. The eggs, which are smooth, hemispherical, and of a very pale green color, are laid singly and hatch in from eleven to thirteen days. The caterpillar feeds on grasses; it is a long time, sometimes several days, in making its exit from the shell, which it then devours and next proceeds to make a rude nest near the joint of a blade of grass by drawing the edges nearly together by silken threads; if at any time it is at all disturbed, it quits its habitation and makes a new nest, occupying much time in its construction, the edges of the blade being drawn closer and closer by continually shortening threads; when about to change to chrysalis, it forms a tube for its concealment by uniting adjoining grass-blades and lines the cavity closely with silk.

The female of this species is dimorphic, one form resembling the male in color, the other (pocahontas) melanic, all the darker markings being extended and the brighter ones obscured.

FAMILY SKIPPERS. 169

Another species of this genus, *A. logan*, a southern form, is found over nearly the same parts of our district as *A. zabulon*, but is far less abundant, though it is not uncommon in the West and especially beyond the Mississippi ; and another species, found in New Jersey and described under the name of *Pamphila aaroni*, is said to be closely allied to these two species and may belong in the same genus.

56. Genus Erynnis.

ERYNNIS SÁSSACUS—THE INDIAN HESPERID.
(Hesperia sassacus, Pamphila sassacus.)

Butterfly.—Upper surface of wings tawny, the outer margin of the fore wings and all the margins of the hind wings heavily bordered with blackish brown, the bordering of the fore wings indented beyond the cell as if to receive the dark longitudinal patch lying just outside it ; discal dash of the male velvety black, slender, slightly arcuate, tapering a little at each end. Under surface pale greenish buff, the markings of the fore wings obscurely traced, and beyond the middle of the hind wings a faint bent row of five not very large, square, pallid spots. Expanse about $1\frac{2}{3}$ inches.

Caterpillar and **Chrysalis** undescribed.

This butterfly is found everywhere in the southern half of our district in fields and meadows. It is single-brooded and probably winters as a chrysalis. The butterfly appears about the last of May and disappears by the middle of July. The eggs, which are smooth, hemispherical, and almost chalk-white when laid, become dirty yellow afterwards; they are laid singly and hatch in about twelve or fifteen days. The caterpillar is very plump at birth and feeds on grasses,—Panicum and doubtless others; it is very sluggish and less cleanly than others of the tribe and makes, at least at first, scarcely an apology for a nest, living near the joints of grasses where the blade embraces the stem.

Several other species of this genus are found in our district : *E. manitoba*, sparingly in its northernmost limits ; *E. metea*, known only in a few localities in southern New England and in Wisconsin ; *E. attalus*, a southern species occasionally occurring in our southern borders ; and *E. uncas*, which has been taken in Pennsylvania and extends to Colorado.

57. Genus Anthomaster.

ANTHOMÁSTER LEONÁRDUS—LEONARD'S HESPERID.

(Pamphila leonardus.)

Butterfly.—Upper surface of wings dark brown, the fore wings with an extramesial series of tawny spots, all but the uppermost large ; discal dash of male black, largest and arcuate at base, very long and slender ; hind wings with a moderately broad extramesial pale tawny band, crossed by dark nervures. Under surface cinnamoneous, the markings of the upper side repeated but paler, on the hind wings white and the band narrowed, lengthened, and more definite. Expanse more than 1½ inches.

Caterpillar.—Head black. Body naked, briefly pilose, pale green dotted with black, the thoracic shield fuscous with black margins (immature ; full-grown caterpillar unknown).

Chrysalis.—Unknown.

Found throughout most or all of our district in open country, but unrecorded from Minnesota and Wisconsin, eastern Maine and eastward. It hibernates as a partly-grown caterpillar, possibly before moulting, and is single-brooded, flying at the end of August and in September. The eggs, which are high hemispherical, smooth and white, are laid upon the blades of the food-plant singly and hatch in from fifteen to twenty days. The caterpillar feeds upon Agrostis and doubtless other grasses, wandering about the blades in the autumn and constructing then no nest of any kind.

58. Genus Polites.

POLÌTES PÉCKIUS—THE YELLOW-SPOT.

(Pamphila peckius, Hesperia wamsutta.)

Butterfly.—Upper surface of wings dark brown, marked with tawny in an extramesial series of elongate spots, reduced to dots and removed outwardly beyond the cell of the fore wings, and crossing but half of the hind wings ; discal dash of male velvety black, sinuous and interrupted before the middle. Under surface cinnamoneous, the markings of the fore wings repeated in yellow, on the hind wings consisting of a very large and very irregular polypoid patch of lively yellow, made up of an oblique basal and a very broad transverse extramesial band which is abruptly broadened in the middle and thus blends with the basal band. Expanse 1¼ inches.

Caterpillar.—Head piceous, rugulose. Body naked, briefly pilose, pale brown, thickly dotted with inky black, giving the whole a griseous appearance; a blackish dorsal line; thoracic shield broad and black (immature; full-grown caterpillar unknown).

Chrysalis.—Unknown.

Found everywhere in our district in open country, and one of our commonest butterflies. It probably hibernates either as a full-grown caterpillar or as a chrysalis; it is single-brooded in the northernmost parts of our district, flying from the last of June to the middle of August, while in the other portions it is double-brooded, flying first from the end of May to the middle of July or later, and again in August and September. The eggs, which are smooth, hemispherical, at first white with a greenish tinge, afterwards decorated with coarse red dendritic markings, are laid singly and hatch in from ten to fifteen days according to the season. The caterpillar feeds on grasses and is very uneasy, roaming about a great deal, making very slight and delicate nests, otherwise similar to those of its allies, and is easily alarmed.

59. Genus Thymelicus.

THYMÉLICUS MÝSTIC—THE LONG-DASH.

(Hesperia mystic, Pamphila mystic.)

Butterfly.—Upper surface of fore wings tawny, brightest in the male, with a very broad outer margin of dark brown and two large dark patches, one just beyond the tip of the cell, the other beneath it at the base; discal dash of male very slender, slightly arcuate, blackish brown, followed below by a rather large, rounded, soft brown patch: hind wings dark brown with an equal, short, extramesial tawny band and a tawny spot at base.

Under surface orange buff (male) or tawny cinnamoneous (female), often infuscated, the brighter markings of the upper surface vaguely repeated and paler, the band of the hind wings generally indistinct in the male. Expanse 1¼ inches.

Caterpillar.—Head reddish brown. Body naked, briefly pilose, dull brownish green, sprinkled with darker dots and having a dark dorsal line; thoracic shield brownish black, in front of it dirty white. Length 1 inch.

Chrysalis.—Unknown.

This butterfly is undoubtedly found over the whole of our district, though it is recorded from few localities in the West; it frequents open grassy fields, and hibernates as a caterpillar; it appears to be single-brooded in the northernmost parts of its range, flying toward the end of June; but over most of our district it is double-brooded, first appearing very early in June or even late in May and rarely flying into July, and being again on the wing from the middle of July to September; but probably in somewhat scantier numbers, for some of the caterpillars of the first brood, though full fed, have not changed to chrysalis when winter appears, when the caterpillars of the second brood are partly grown. The eggs are smooth, hemispherical, and very pale green, are laid singly very lightly affixed to grass-blades, and hatch in from eight to fourteen days, according to place and season. The caterpillar feeds on grasses, does not devour its forsaken egg-shell, and makes a tubular nest of grass-blades, to which it retires on the slightest alarm; it is firmly constructed of many blades and many threads and the interstices covered with a gauze-like open framework.

———

Other species of this genus found in our district are *T. aetna*, a southern species not very uncommon as far north as Canada; and *T. brettus*, known mostly from the southern coast, but extending northward into Connecticut, and reported also from Wisconsin.

FAMILY SKIPPERS.

60. Genus Limochores.

LIMOCHORES TAUMAS—THE TAWNY-EDGED SKIPPER.

(Pamphila cernes, Hesperia ahaton.)

Butterfly.—Upper surface of wings dark brown, the fore wings with a large costal bright tawny patch (male), or an obscure tawny streak along outer half of cell (female), the female with an extramesial series of three upper small yellow dashes and two or three lower large squarish yellow spots, sometimes found indicated in the male; discal dash of male black, sinuous, heavy. Under surface rather dark brown, flecked uniformly on hind wings with greenish yellow giving a grayish olivaceous effect, the lighter markings of fore wings repeated. Expanse scarcely $1\frac{1}{4}$ inches.

Caterpillar.—Head black, coarsely punctured. Body naked, briefly pilose, rich purplish brown with a green tinge, finely mottled with gray and dark purplish brown; first thoracic segment milk-white above, the shield piceous. Length 1 inch.

Chrysalis.—Light brown with slight and delicate infuscations, the thorax darker, the head black, the whole dotted sparsely with fusco-ferruginous; surface vermiculate; tongue reaching the eighth abdominal segment. Length fully $\frac{1}{2}$ inch.

Everywhere a common insect in open fields. It hibernates in the chrysalis and is single-brooded in the northernmost parts of our district, flying late in June and in July; but double-brooded over most of it, the first brood appearing the last week in May, abundant in June, and seen in scanty numbers all through July; the second brood, less abundant than the first (probably because some chrysalids of the first brood winter over), appearing pretty early in August and flying through September. The eggs, which are smooth, hemispherical, and pale green, are attached lightly and singly to grass-blades and hatch in from eleven to fifteen days. The caterpillars feed upon grasses, such as Panicum and Triticum, and are indolent, passive, and timorous, feeding only by day, rarely leaving their nests and then going but a little distance. For change to chrysalis they make a light, nearly erect cocoon about an inch long by catching a few blades of grass together and lining them with silk.

Other species of this genus found in our territory are *L. bimacula* and *L. manataaqua*, both found throughout its southern half and tolerably common; *L. pontiac*, found in the same places but much rarer, commoner in the West than in the East; and *L. palatka*, found only in the West—Nebraska, Illinois, and Indiana—and little known.

A number of other genera of the Smaller Skippers are found in our district, some of them not uncommonly, but they are mostly obscure forms and their distribution imperfectly known, and they have therefore been omitted from consideration. Such are **Oarisma**, with one species, *O. powesheik*, a western form found in northern Illinois, Iowa, Nebraska, and westward; **Potanthus**, represented by *P. omaha*, known only from West Virginia and Colorado; **Pamphila**, a highly interesting type with one species, *A. mandan*, found in the high north and invading our northern border; **Amblyscirtes**, with two species, *A. vialis*, found sparingly over all our region, and *A. samoset*, known mostly from New England but also from as far west as Iowa and south as Georgia; **Poanes**, with a single conspicuously marked species, *P. massasoit*, occurring here and there in the southern half of our district; **Phycanassa**, with one species, *P. viator*, a southern form which has once or twice occurred far north at widely separated localities; **Hylephila**, represented by *H. phylaeus*, a very abundant southern type which occasionally invades our southern borders, even as far as southern New England; **Atalopedes**, with one species, *A. huron*, a southern form reaching northward over half of our district; **Euphyes**, with three species: *E. metacomet*, found over all but the extreme eastern part of our district and sometimes pretty common; *E. verna*, which ranges nearly as far and is rarer; and *E. osyka*, a southern species which has been taken in northern Indiana; **Lerodea**, one species of which, *L. fusca*, a southern form, is said to be common about Philadelphia, Penn.; **Prenes**, with two species, *P. ocola* and *P. panoquin*, both southern types but occasionally taken in our district, the former in Indiana and Pennsylvania, the latter in New Jersey; **Calpodes**, with one species, *C. ethlius*, a tropical form which has been once taken in New York; **Oligoria**, represented by *O. maculata*, a southern type also once taken in New York; and finally **Lerema**, represented by two species, *L. accius*, a southern coast species occurring rarely as far north as Massachusetts, and *L. hianna*, which has been found in scanty numbers from Massachusetts to Nebraska.

EXPLANATION OF SOME TERMS.

Other words are explained by the context.

Acutangulate: forming less than a right angle.
Anal angle (of the wing): see Figure, p. 60.
Antennae (of the butterfly): the two long slender rods projecting from the top of the head.
Armature (of the legs): the corneous attachments or appendages, spines, claws, etc.
Atavistic: pointing backward to ancestry.
Bifurcate: with two prongs.
Blind (said of ocelli on wings): with no pupil.
Border and Margin are used interchangeably.
Cell, or Discoidal cell: see Figure, p. 60.
Coronal: at the summit.
Corneous: of a horny texture.
Costa or Costal margin: see Figure, p. 60.
Costal vein: see Figure, p. 60.
Crenate: wavy or scalloped.
Crenulate: the same, but to a less degree.
Cycle: regularly recurring series.
Denticulate: covered with tooth-like points, or with a toothed margin.
Dimorphic: appearing under two distinct forms.
Discal dash or stigma: a small spot (peculiar to the male of some Hair-Streaks and Skippers) on the fore wings, at the end of the cell.
Discoidal cell: see Figure, p. 60.
Disk: central portion of the wing.
Dorsal shield (of the caterpillar): the thickened plate on top of first thoracic segment.
Emargination: a notch or rounded excision.

Entire (of a margin): whole and even.
Environment: surroundings and their influence.
Eversible: capable of being turned inside out.
Extramesial: beyond the middle.
Falcate: sickle-shaped, convex on one side, concave on the other.
Fenestrate: resembling a window or opening.
Frontal triangle (of the caterpillar): the large triangular piece on the face.
Granulated: covered with small, grain-like elevations.
Hemisphere (of the caterpillar): one lateral half of the head.
Hibernaculum: wintering nest of the caterpillar.
Incisures: impressed lines, separating the segments of the body.
Infralateral: just below the lateral line or a line midway between the middle of the back and the spiracles.
Infrastigmatal: just below the spiracles, or the spiracle-line.
Inner margin (of the wing): see Figure, p. 60.
Intergrades: forms intermediate between others.
Internal vein: see Figure, p. 60.
Interspace: space between two adjoining nervules.
Intramesial: before the middle.
Irrorate: bedewed or uniformly sprinkled.
Isotherm: line of equal temperature.
Lateral (of the caterpillar): along a line midway between the middle of the back and the spiracles. Sometimes applied loosely to the sides in general.
Laterodorsal: situated midway between the lateral and mediodorsal (which see).
Lunulate: in the form of lunules or moon-shaped crescents.
Mandibles (of the caterpillar): the biting jaws.
Margin and Border are used interchangeably.
Median vein: see Figure, p. 60.

Mediodorsal: lying along the middle line of the back.
Mesial (of the wing): along the middle.
Obsolete: very nearly or quite wanting.
Ocellar tubercles (of the chrysalis): the prominences arising from the region of the eyes.
Ocelli (of the caterpillar): the simple eyes, each composed of a single facet.
Ocelli (of the wing): eye-like spots.
Onisciform: shaped like a wood-louse (Oniscus), or slug-shaped, i.e., flattened beneath and more or less ovate in outline.
Outer angle (of the fore wing): the angle at the lower limit of the outer margin.
Outer margin: see Figure, p. 60.
Papillae: small, pimple-like elevations.
Papillate: covered with papillae.
Parenchyma: the softer cellular tissue of a leaf.
Pilose: covered with a nap of short hairs.
Polymorphic: appearing under many different forms.
Polyphagous: feeding on many different plants, omnivorous.
Prebasal (of the wing): next but not at the base.
Precostal vein: see Figure, p. 60.
Premarginal: just before the margin (especially outer margin).
Process: any projecting appendage or part.
Produced: extended.
Rectangulate: forming a right angle.
Saddle (of the chrysalis): the depressed part of the back at the base of the abdomen.
Shield: see Dorsal shield.
Stigma: see Discal dash or stigma.
Stigmata: spiracles or breathing-pores.
Stigmatal: along the line of the spiracles.
Sub- (as a prefix) signifies nearly, as subglobular = nearly globular.

Subcostal vein: see Figure, p. 60.
Submarginal: next to but not on the margin; usually applied to the outer margin.
Submedian vein: see Figure, p. 60.
Subobsolete: present, but faint, nearly obsolete.
Supralateral: just above the lateral line, or a line midway between the middle of the back and the spiracles.
Tectate: inclined obliquely on opposite sides, like the roof of a tent.
Thoracic shield: see Dorsal shield.
Tiarate: shaped like a turban.
Trimorphic: appearing under three distinct forms.
Tubercles: see Wing-tubercles.
Vermiculate: resembling interlacing worm-tracks.
Wing-tubercles (of the chrysalis): elevations at the base of the wing-cases; the front one, when there are two, is distinguished as the basal wing-tubercle.

APPENDIX.

INSTRUCTIONS FOR COLLECTING, REARING, PRESERVING, AND STUDYING.

(From the author's "Butterflies, their Structure," etc.; with slight changes.)

HAPPILY the time is past when butterfly-collectors devote their entire attention to the perfect insect. They at least rear them from the caterpillar or chrysalis to obtain fresher and more beautiful specimens for their cabinets; and it is to be hoped that any young enthusiasts who may use this book will be quite as ready to collect, preserve, and study the earlier stages as the full-grown insect. It therefore needs no apology from me in giving here more space to instructions concerning the pursuit of the immature than of the mature form.

The best method of raising butterflies is to obtain eggs from the parent and rear them to maturity. This is by no means difficult and is full of interest; it is only necessary to know the food-plant of the caterpillar—and that of nearly all our northern species is ascertained; or if it is not known, it may often be inferred from that of neighboring species, or discovered by patiently following the female as she flits from leaf to leaf, and noticing the plants she chooses whereon to lay her eggs. The butterfly generally selects the middle of the day for this duty, but the eager youth must not expect at once to obtain her secret, for he will find himself only too often foiled. Once known, the way is comparatively easy ; catch a female, selecting for the purpose one which has evidently been flying for at least a few days, and which is gravid with eggs, and inclose her beneath a gauze covering upon the growing plant. If it be a tree or bush, tie a bag of mosquito-netting over a bough, taking care that there are some tender leaves upon it (and no ants), and so arrange the bag that the butterfly may rest naturally upon them ; inclose the butterfly and she will pretty certainly deposit eggs in the course of a day or two. Or, if the plant be one of small size, use a headless keg, covered at one end with gauze ; even a discarded vegetable-can will serve the purpose ; or again, a canopy can be made over a plant by thrusting the ends of a couple of bent twigs into the ground and covering with gauze. A bit of sugared apple or other fruit should be inclosed as food.

After a few days' confinement the prisoner should be set free. If she has not then laid eggs, she probably cannot, and she should be released. If she has yielded the desired harvest, she should be rewarded with liberty. When obtained, the leaves or twigs upon which the eggs are found may either be left where they are or carried home to more convenient quarters.

It is not easy to preserve eggs entire. If they do not hatch they are apt to shrivel, excepting such as have a dense pellicle, like the hemispherical eggs of the smaller skippers or the tiarate eggs of the blues and coppers; it is nearly impossible, too, to prick the egg and save its form. The best way is to watch for the egress of the caterpillar and the moment it is free separate it from the shell, which it will otherwise devour; in that way I have obtained a considerable collection of these little gems. Or they may be obtained from the plants on which they have been laid naturally, by searching the food-plants carefully; they are not so difficult to detect as might be supposed; many of these will be found attacked by minute parasites, which generally make their exit through a single minute hole, leaving the egg in an admirable condition for the cabinet. The eggs can then be gummed, with or without the leaf on which they are laid, upon triangular bits of card-board, pinned and transferred to the cabinet. Inspissated ox-gall, diluted with an equal quantity of thick gum arabic, makes the best material for attachment to the card.

In rearing from the egg the greatest difficulty is during early life; young caterpillars must have the freshest and tenderest food and not too much confinement. With all precautions many will be lost, for they are so small that it is difficult to keep track of them, and some are very prone to wander when their food does not suit them. Some open vessel with the growing plant is the best receptacle; in place of this a similar vessel (the larger the better) holding moist sand in which a sprig of the food-plant is plunged may be used—covered if convenient with gauze to prevent the escape of the caterpillar. The vessel should be placed in the light, but not in the sun, and for many kinds it is well to lay chips or bits of bark upon the ground, beneath which the caterpillars may hide. At each moult the caterpillar remains motionless, refusing to feed for twenty-four hours or more, and at such times it should not be disturbed. It is best never to touch them, and, when necessary to change the food, the old leaf with the caterpillar upon it should be put aside or upon the fresh food, and only removed when deserted by the caterpillar. When older the creature will bear rougher treatment and may often be confined in a nearly tight tin or earthen vessel with freshly-plucked leaves; but all caterpillars will not bear this treatment, and care should always be taken that their quarters do not become in the least foul.

A very convenient form of breeding-cage or vivarium is shown in Fig. 2, and is thus described by Mr. Riley:

FIG. 2.—Breeding-cage, described in the text.

"It comprises three distinct parts: first the bottom board (a), consisting of a square piece of inch-thick walnut with a rectangular zinc pan (ff) four inches deep fastened to it above, to prevent cracking or warping, facilitate lifting, and allow the air to pass underneath the cage. Second, a box (b), with three glass sides and a glass door in front, to fit over the zinc pan. Third, a cap (c) which fits closely to the box, and has a top of fine wire gauze.

To the centre of the zinc pan is soldered a zinc tube (d) just large enough to contain an ordinary quinine bottle. The zinc pan is filled with clean sifted earth or sand (e), and the quinine bottle is for the reception of the food-plant. The cage admits of abundant light and air, and also of the easy removal of excrement and frass which falls to the ground; while the insects in transforming attach themselves to the sides or the cap according to their habits. The most convenient dimensions I find to be twelve inches square and eighteen inches high; the cap and the door fit closely by means of rabbets, and the former has a depth of about four inches to admit of the largest cocoon being spun in it without touching the box on which it rests. The zinc pan might be made six or eight inches deep, and the lower half filled with sand, so as to keep the whole moist for a greater length of time. A dozen such cages will furnish room for the annual breeding of a great number of species, as several having different habits and appearance, and which there is no danger of confounding, may be simultaneously fed in the same cage."

The best success will always attend efforts to place the prisoner in conditions as nearly natural as possible; but in rearing out-of-doors it is more difficult to keep track of your charges, and they are of course more subject to their natural enemies, which are numerous and vigilant. Moreover it is then nearly impossible to obtain the cast-off heads of each moult, which are well to preserve for comparative study at leisure, or to complete the tangible marks of the life-history of the insect.

Such caterpillars as construct nests in which to live when not feeding, and especially such as then live a great while in the caterpillar state, as for instance nearly all the skippers, are the hardest to rear satisfactorily apart from their natural homes; they do not like to live in a dried-up house, nor to be continually wasting their energies in the construction of new ones, so that one's ingenuity is often taxed to keep them happy; but patience and careful attention to their natural conditions will reap their reward, and I believe it is possible with care to breed any of our species in confinement. Caterpillars found partly grown in a state of nature may be reared in confinement for the rest of their lives with equal ease ; only one labors then under the disadvantage, if he cares only for the butterfly, of being rewarded for his pains merely by a fine batch of minute hymenopterous parasites or a bristling fly or two. To one, however, who is interested in the entire history of these creatures, this is not altogether a loss, for he will add perchance to his stock of butterfly parasites, of which for some species many different kinds are already known.

The search for caterpillars in their haunts is often very easy, especially if their food plant, habits, and seasons are known ; to search for a caterpillar out of season is an anachronism one will not enjoy. Partly-eaten leaves are one of the best guides to the discovery of caterpillars; while such as construct nests of any sort are very readily detected, especially when the nests are so built as to expose the under surfaces of leaves, where their upper surfaces would be expected, as in the case of many of the higher skippers. The caterpillars of the blues, coppers, etc., are perhaps the most difficult to find, because they so nearly resemble in color the surfaces on which they rest ; the same is true of the caterpillar of our common yellow butterfly ; but when one has once discovered them, and knows *how they look* in their natural situations, the search becomes much easier. Others again feed mostly by night and retire by day to the covert of dead leaves on the ground or beneath sticks, and must be sought by the aid of the lantern. Such in particular are the caterpillars of our satyrs and fritillaries.

Some caterpillars pass the winter in that state, either just hatched, half grown, or nearly mature. To keep these safely through our long winter and prevent their recovering from their dormancy before food for them can be obtained in the spring is one of the most difficult tasks. It is best, as a general rule, to place them in closed or nearly closed vessels, not too small, in a dry but cool cellar, and not to move them until their food-plant is again in leaf. Mr. Edwards has succeeded well with some of those which have eaten little or nothing before going into winter quarters, by placing them through the winter in an ice-house, which would seem to be rather heroic treatment at first sight ; but in almost any other situation they are liable to rouse from their lethargy too early in the spring, the critical period, no doubt, of their life. For collecting caterpillars, pocket tin boxes are the best receptacles.

The satisfactory preservation of the caterpillar for the cabinet is far easier than is generally supposed. For anatomical purposes it is much better to dissect fresh specimens, but very much may be done with specimens that have been preserved in not too strong alcohol, or in glycerine and carbolic acid. For the study of the markings or of the external features or form, nothing equals the method known as inflation, where only the pellicle and its appendages are preserved, and which has the advantage of allowing the caterpillar to be readily placed in an ordinary cabinet beside the other forms of the creature's life; also of preserving in their natural relations all the spines and hairs which clothe the body, and of allowing these to be studied at pleasure; specimens preserved in any fluid, on the contrary, are difficult to handle conveniently, and their examination is unsatisfactory from the matting of the hairs and spines.

The instruments necessary for inflating are a small tin

oven, a spirit-lamp, forceps, a pair of finely-pointed scissors, a bit of rag, a little fine wire, and a wheat straw, or a glass tube drawn to a fine point. The oven is simply an oblong tin box, about 2¼ inches high, 2½ inches wide, and 5 inches long; the cover is of glass, and one end of the box is perforated by a circular hole 1¼ inches in diameter.

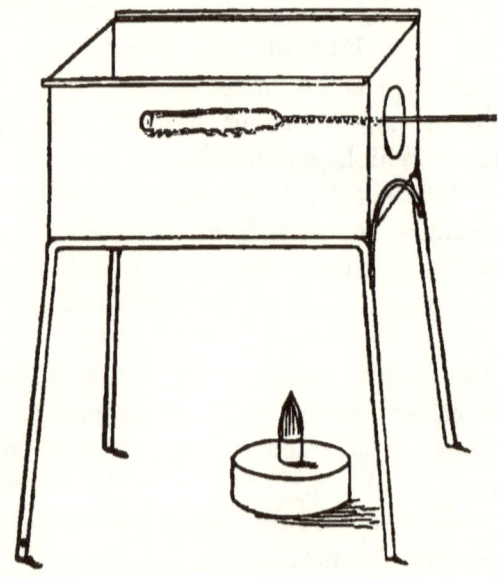

FIG. 3.—Oven and lamp for preparing caterpillars by inflation.

The oven rests upon a wire standard as in the woodcut [Fig. 3]. No soldering should be used upon the oven, as it would soon be melted. The wire for the caterpillar should be very fine and annealed; the best is that wound with green thread and used for artificial flowers. It should not be more than half a millimetre in diameter. [Fig. 4.]

Kill the subject by a drop of ether or by a plunge in spirits. Then placing the caterpillar in the left hand, so as to expose its hinder extremity beyond the gently closed thumb and first two fingers, enlarge the vent slightly at the lower edge by a vertical cut with the scissors; next lay the larva either upon bibulous paper on the table, or upon soft cotton cloth held in the left hand, and press the extremity of the body with one finger, always with the interposition of cloth or paper, so as to force out some of the contents of the body; this process is continued from points successively farther back, a slight additional portion of the contents of the body being gently pressed out with each new movement. Throughout all this process great care should be taken lest the skin should be abraded by too violent pressure, and lest any of the contents of the body soil its exterior or become entangled in the hairs or spines; to avoid the latter, the caterpillar should be frequently removed to a clean part of the cloth. When a portion of the intestinal tube itself becomes extruded, it should be gripped with a pair of strong forceps, and, the head remaining in the secure hold of the left hand, the tube should be forcibly but steadily torn from its attachments; with this most of the contents of the body will be withdrawn, and a delicate pressure passing with a rolling motion from the head toward the tail will reduce the subject to a mere pellicle.

FIG. 4.—Wound wire for supporting caterpillars, × 20.

The alcohol-lamp is now lighted and placed in position beneath the oven; a wheat straw is selected, of the proper size to enter the enlarged vent, and the tip, after being cut diagonally with sharp scissors or a knife, is moistened a little in the mouth (to prevent too great adhesion of the skin to the straw) and carefully introduced into the opening of the caterpillar; the process may be aided by blowing gently through the straw. When the skin is slipped upon all sides of the straw to the distance of about a fifth of an inch, without any folding of the skin and so that both the anal prolegs protrude, a short delicate pin (Edelston and Williams, No. 19, is best) is passed through the anal plate and the straw. If a glass tube is used, the anal plate must be fastened to it by winding with silk.

By this time the oven will be sufficiently heated to begin the drying process, which consists simply in keeping the caterpillar in the oven, extended horizontally by blowing gently and steadily through the straw, as one uses a blowpipe. Too forcible inflation will make the caterpillar unsightly by distending unnaturally any spot that may have been weakened or bruised in the previous operation; the caterpillar should be kept slowly but constantly turning, and no harm will result from withdrawing the creature from the oven and allowing it to collapse, to gain breath or rest; only this relaxation should be very brief. The caterpillar should be first introduced into the oven while inflated by the breath, and so placed that the hinder extremity shall be in the hottest part, directly above the flame, for it is essential that the animal should dry from behind forward; yet not altogether, for as soon as the hinder part has begun to stiffen (which can readily be detected by withholding the breath for a moment) the portion next in front should receive partial attention, and the caterpillar moved backward and forward, round and round over the flame. During this process any tendency of the caterpillar to assume unnatural positions may be corrected—at least in part—by withdrawing it from the oven and manipulating it; during inflation, the parts about the head should be the last to dry and should be kept over the flame until a rather forcible touch will not cause it to bend.

To secure the best results, it is essential that the oven should not be too hot; the flame should not be more than an inch high, and its tip should be one or two inches from the bottom of the oven.

When the skin of the caterpillar will yield at no point, it is ready for mounting. The pin is taken from the straw, and the caterpillar skin, which often adheres to the straw, must be gently removed with some delicate, blunt instrument, or with the finger-nail.

A piece of wire a little more than twice the length of the caterpillar is next cut, and, by means of forceps, bent as in Fig. 5, the tips a little incurved, a little shellac* is

Fig. 5.—Wire bent into shape to insert into the caterpillar; not enlarged.

placed at the distal extremity of the loop, the wire is held by the forceps so as to prevent the free ends of the wire from spreading, and they are introduced into the empty body of the caterpillar as far as the forceps will allow; holding the loop and removing the forceps, the caterpillar is now pushed over the wire with extreme care, until the hinder extremity has passed half-way over the loop, and the shellac has smeared the interior sufficiently to hold the caterpillar in place when dry; the extremities of the parted wires should reach nearly to the head. Nothing remains but to curve the doubled end of the wire tightly around a pin with a pair of strong forceps and to place the specimen, properly labelled, in a place where it can dry thoroughly for several days before removal to the cabinet.

For more careful preservation and readier handling, each specimen may be placed in a glass tube, like the test-tube of the chemist. The wire is then first bent in the middle and the bent end inserted in a hole bored in the smaller end of a cork of suitable size, so as nearly to pass through it; the loops are then formed as above; both ends

---

* To prepare this, the sheets of dark shellac should be preferred to the light, and dissolved in forty per cent alcohol.

of the cork are varnished, and a label pasted around the portion of the cork which enters the tube, thus guarding both specimen and label from dust, and the latter from loss or misplacement. After two or three days the cork with the caterpillar attached is placed in its corresponding tube, and the tube may be freely handled.

Modifications of this system will occur to every one. Dr. Gemminger uses a syringe for the extraction of the contents as well as for the inflation of the emptied skin. For an oven, the Vienna entomologists employ an ordinary gas-chimney, open at both ends and inserted in a sand-bath, which prevents, perhaps, the danger of too great heat.

In rearing caterpillars for the after-stages, care must always be taken to provide in season a suitable place in the breeding-cage for the chrysalis to suspend itself: a twig for such as prefer such situations; a bit of shingle near the top of the cage for those that suspend themselves by the tail, or fasten themselves preferably to flat surfaces; leaves for those that construct some sort of a cocoon. The search for chrysalids in the open air is not likely to meet with great success excepting in a few instances, such as the imported cabbage butterfly, whose chrysalids can be found in only too great abundance beneath palings or on the under edge of clapboards on farm-houses; those of the blues and their allies may often be found beneath stones, but one must be an enthusiast to follow the search at all successfully; such as fall into the hands of the general entomologist must be counted as clear gain; yet these will often repay him who studies also the parasites of butterflies, so often are they found to be infested.

The preservation of chrysalids with their colors is easy for all that are not of some green tint; and these are few. Long-lived chrysalids are not easily killed excepting by extreme dryness. Some will survive a twelve hours' plunge in alcohol, and those that could not would generally lose some of their colors by the immersion. Dry heat is the best method, but it should be accompanied after death by further drying after an opening has been made into the body, lest the contents should decay. Parasitized specimens form the best material for the cabinet, but even shells from which the inmate has escaped can by careful manipulation and a little glue have their separated parts so joined as to answer fairly the desired purpose. Solid specimens can be pinned through one side of the thorax, but the mere pellicle should have the hooks of the tail securely fastened to a little ball of cotton wool or bit of felt, through which the pin may be passed. It is not easy to glue empty chrysalids permanently to cards, and these are very apt to hide the parts one wishes at some future time to examine. Skilful persons may attain some success with thin-skinned chrysalids, like that of the milkweed butterfly, for instance, the shape of which is difficult to retain, by removing the contents through a small opening at one side and stuffing with cotton.

The best form of net for the capture of butterflies is a bag fastened to a hoop or ring of some sort, to which a handle may be attached. The hoop should be made of galvanized iron wire, forming a circle about twelve to fourteen inches in diameter, and the bag, made of double bobbinet and attached to the wire by strong linen or cotton, should taper regularly, have a rounded bottom, and be about thirty inches long, so as to double over the net and and have a few inches to spare. By bending the two ends of the wire as in Fig. 6, they can be dropped into a brass tube and securely fixed in place

FIG. 6. — Net frame for butterflies. *a*, wire ring, with ends bent to insert into the ferrule, *b*; *c*, point where the plug and net handle meet.

by a tight plug of hard wood, leaving the other end of the tube open for the insertion of a removable handle; or a very convenient form of net can be constructed on the following plan shown in Fig. 7 and thus described

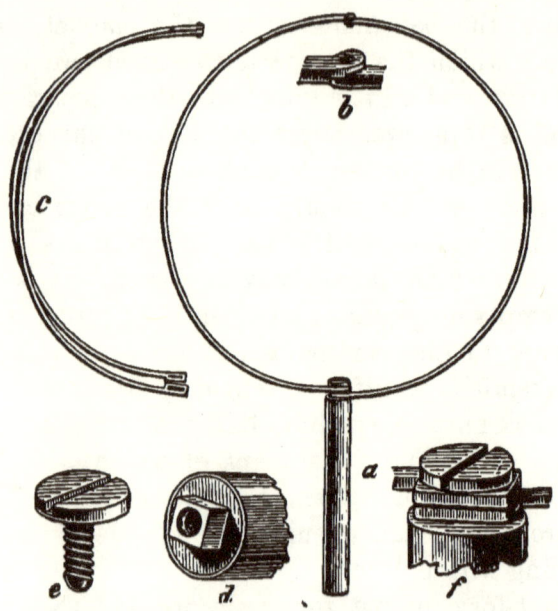

FIG. 7.—Folding net frame, explained in the text.

by Mr. Riley: "Take two pieces of stout wire, each about twenty inches long; bend them half circularly and join at one end by a folding hinge having a check on one side (b). The other ends are bent and beaten into two square sockets (f), which fit to a nut sunk and soldered into one end of a brass tube (d). When so fitted they are secured by a large-headed screw (e), threaded to fit into the nut-socket, and with a groove wide enough to receive the back of a common pocket-knife blade. The wire hoop is easily detached and folded, as at c, for convenient carriage; and the handle may be made of any desired length by cutting a stick and fitting it into the hollow tube *a*, which should be about six inches long." The stick should be about four feet long. Mr. Lintner makes use of a rod with a head [Fig. 8] screwed to one end, in which to fasten an elastic brass ribbon, on which the net is drawn, but which when not in use may be placed inside the hat, while the stick serves as a cane, and the head and bag may be placed in the pocket. An entomologist becomes a less conspicuous personage with such an outfit.

The "chase" for butterflies should rarely be a question of speed; caution and stratagem are better arts; a butterfly should rarely be alarmed, or the game is lost; intent upon a flower, one may even be captured with the fingers by slow approach upon the shady side; many have the habit of returning to a twig they have left, and can be captured by lying in wait near the spot; others will course up and down a roadside, a forest lane, or a hedgerow, and may be easily netted by taking advantage of this habit. Nor should it be forgotten that not a few are very limited indeed in the selection of their haunts, and every kind of spot should be visited; some confine their flight to marshy spots and even to particular bogs; some prefer the open fields; pastures where thistles and other weeds are in flower attract a great crowd; others may be found in openings in the forest where the fire-weed conceals the charred timber beneath its panicles of blue flowers; one will not look in vain upon the goldenrods and blossomed vines which fringe the roadside or stone walls; the shrubbery which loves the margin of slender streams or the edge of thickets is a favorite haunt of many; sheltered valleys with their varying verdure are always a choice resort of the entomologist; but even the tops of rugged mountains or sandy wastes given to sorrel and feeble grasses

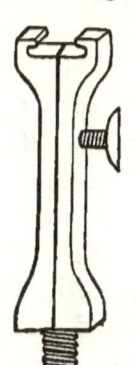

FIG. 8. — Net-head for a removable frame.

will yield their quota; the garden too, the vegetable field, and even the roadside puddles must not be neglected.

One soon learns to capture with a dexterous turn of the net, and no description of the method is worth anything beside a very little experience; when captured the net should be turned to prevent escape and the butterfly gently seized from outside the net, with the wings back to back to prevent its struggling and so bruising itself; it should then be removed to the cyanide bottle, where, especially if placed in the dark pocket, it will soon be motionless, and speedily dies; this is the quickest and easiest mode of death, besides leaving the insect in the most perfect condition. The "cyanide bottle" is simply a phial with a mouth wide enough to readily admit the largest specimens (a smaller size is better for the smaller kinds), into which a little plaster of Paris has been poured over a small lump of cyanide of potassium (a deadly poison, be it noted); or, a lump of cyanide may be inclosed in a piece of chamois-skin wrapped around and tied above the cork, leaving the bottle clean. The cork should be removed only when necessary and for as little time as possible; a season's use will exhaust its best strength even when the utmost care is taken. Some butterflies, especially those having yellow colors, should be left in the bottle only a short time, for they are injured by too long exposure to the vapors, the yellow turning reddish. When removed, on reaching home, or sooner if needed, they should be pinned through the thickest part of the thorax, and in an hour or two, when the fixity of the wings which follows their violent death has passed away, removed to the setting-board.

The best pins for butterflies are Nos. 2, 3, and 4 of Klaeger's make. The setting-board needs no description apart from the figure given [Fig. 9], more than to say that beneath the groove a strip of cork or pith is attached to the board. Bits of glass cut to different sizes answer as well as the card braces represented in the illustration and permit one better to see whether the wing is lying perfectly flat. A needle inserted in a handle is required to move the wings into the desired position, and "to set" the antennæ and legs in a natural attitude; to secure these in the proper place they are supported by insect pins stuck into the board upon one side or the other of the member,

FIG. 9.—Setting-board.

as required. The butterflies should remain upon the setting-board for a fortnight or longer, and placed where they will dry readily but not be exposed to dust. At the expiration of that time they are ready for the cabinet.

When one is away from home conveniences, a very simple device for transportation is to fold oblong bits of paper (rather thin writing-paper is best) into "triangles," as along the dotted lines in this sketch; into this the butterfly is placed, its wings folded back to back and antennae tucked carefully away. The place, date, and circumstances of capture (or a number corresponding to a journal) may be written upon the paper. A great number may thus be packed into a cigar-box or other receptacle, and spread for the cabinet at leisure, months or even years after collection. For this purpose moistening-pans are needed. A glass or stoneware dish is the best, the top ground so as to allow a sheet of glass to cover it perfectly; upon the bottom moistened sand is placed, covered by fine brass wire netting. A few papers with their inclosed butterflies are placed in it, and the cover left on for twenty-four hours or thereabouts, when the insects may be handled nearly as if just caught.

Damp, grease, and museum pests are the great destroyers of insect collections. To avoid the first, one has only to see that his cabinet is in a dry place, with a play of air around it. To avoid grease, insects should be thoroughly dried before being admitted to the cabinet, and all use of cedar wood in constructing the latter should be avoided; benzine is perhaps the best material for removing it. Against museum pests one can be safe only by a constant, vigilant, searching oversight of his collection, or the use of boxes which they cannot enter; even then care must be taken not to introduce them one's self by placing infested specimens in the collection: for this purpose it is well to establish a safe quarantine.

For a permanent cabinet nothing can excel the drawers made after the Deyrolle model, now in use by the Boston Society of Natural History. I have tried them for many years and find them entirely pest-proof. They are made [Fig. 10], with a cover of glass set in a frame which is grooved along the lower edge, and thus fits tightly into a narrow strip of zinc, set edgewise into a corresponding groove in the drawer; the grooves beyond the point of intersection of two sides are filled with a bit of wood firmly glued in place. It is hardly necessary to say that the sides of the drawer and the frame of the cover should be made of hard wood; soft wood would not retain the zinc strip. The zinc should be perfectly straight and the ends well matched; if this be done, nothing can enter the box when it is closed. The bottom should set in a groove in the sides and not be flush with their lower edge, so that the drawer may slide easily. A similar box with a wooden rabbet is used at the Museum of Comparative Zoology at Cambridge; but it cannot possibly be so tight, and re-

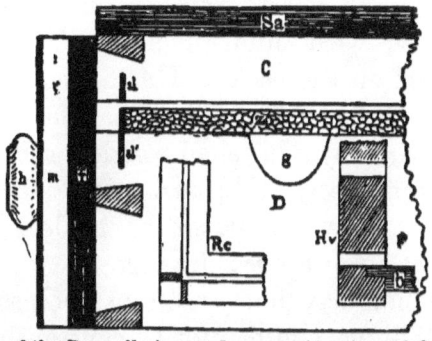

Fig. 10.—Model of the Deyrolle insect-drawer, side view of front end, with the cover raised. D, bottom of drawer; C, cover of same, raised a little; *f*, front piece, with moulding (*m*) and handle (*h*), glued to bottom piece; *sa*, sash; *sl*, slit in cover into which the zinc strip (*z*) fits; *sl'*, slit in bottom, into which it is fastened; *g*, bevelled groove, to allow the finger to raise the cover: *Hv*, hind view of one end of the bottom to show the insertion of the bottom (*b*); *Rc*, reverse of one corner of cover to show the grooves filled beyond their junction. All the figures half size.

quires hooks on the sides to keep the cover down; it has the advantage of greater cheapness, as it can be made of soft wood, but is at the same time clumsier. My own drawers are made of cherry sides, and have also a false front attached to them, furnished with mouldings and handles so as to present a not inelegant appearance; and, exclusive of the cork with which they are lined, cost $2.65 each; they measure inside 18¾ inches long, 14 inches wide. and 1⅞ inches deep, not including the cork lining.

It is best always to cover the bottom of such drawers with cork or pith wood or similar soft substance, as it is difficult both to insert and to withdraw the pins readily in any ordinary wood, however soft; and the sides and bottom should afterwards be covered with thin white paper for neatness' sake.

Drawers like these are rather large for small collections, but any smaller size is wasteful of space for arranging the larger species of wide expanse of wing. Some, however, still prefer smaller sizes for convenience of study, and use boxes shaped like a quarto volume, the cover hinged and the whole lined with binder's cloth. The volumes can then be lettered on the back and arranged as in a library, and certainly have a neat appearance. Such books can be made safer either by a bevelled wooden rabbet where the top and bottom meet, or by arranging within a second glass cover, but they can never be made so fully proof against pests as an unhinged drawer.

A very common box, but unsafe as soon as a collection becomes at all large and cannot be constantly watched in every part, is a simple wooden box nine by fourteen inches in size, in which both top and bottom, made separate, are put to use by being lined with cork. In this case the box must, of course, be much deeper. Such cases can be made in numbers for fifty cents each, exclusive of the cork, and answer very well for beginners, but will be discarded after a time if the collection increases, unless the owner has sufficient leisure and patience to watch his treasures carefully.

The best way to begin the study of butterflies is to attempt to follow out the life-history, write the biography, in short, of every kind found in one's own neighborhood. No one place will yield much above one hundred species, and, if the rarer kinds be omitted, not nearly so many. Yet any one who will accomplish this will add materially to what is known, and he will find his way pleasanter, his occupation more fascinating at every step. He need be provided at the outset with a very moderate stock of the articles mentioned in the preceding pages. He should keep a journal devoted exclusively to a record of his daily notes, which will prove more and more useful in each succeeding year. Beginning with the eggs laid by imprisoned females or found in the open field, he should note every change which transpires, describe, and, if possible, *draw in detail* every stage, giving to each separate lot a distinctive number, which it should keep until its name is known. As his stock enlarges and his knowledge increases, comparative study will supersede many of his earlier descriptions; but these will not have been without their value; they will have cost no more than they are worth; his knowledge will have been gained through, as well as at the expense of, his earlier work, none of which will he regret; he can therefore be neither too minute nor too exact, nor can he afford to relax any endeavor until he has proved it unnecessary.

He should preserve in his permanent collection specimens to illustrate every condition of the creature's life, as well as all objects which illustrate its habits and vicissitudes. Especially should all variations be observed. The egg with the leaf upon which it is laid in a state of nature; not only the caterpillar at every stage, but in all the attitudes it assumes, the nests it weaves, the half-devoured leaves to show its manner of feeding, the ejectamenta, the parasites by which it is beset; not only the chrysalis, but the emptied skin; the butterflies of each brood, together with some preserved in their natural attitudes when at rest, and when asleep; and such dissections of the external parts as can be separately mounted and cannot otherwise be readily seen; also the wings and body of the butterfly denuded of their scales, to study the structural framework of the insect; and, when possible, dissections of the internal parts preserved in alcohol.

Every pinned specimen, excepting such as illustrate the anatomy only, should bear upon the pin a label giving the place and date of capture, and, when necessary, a number referring to a catalogue or note-book in which memoranda may be entered to any extent that is desired. The name of the species may be given on a separate label at the head of each collection of objects which illustrate its history; and the name may, of course, also be added at will to any specimens which, once determined, may require redetermination if misplaced and not specially marked.

In rearing it is essential that every breeding cage or pot should be marked with a number or by other means to indicate its contents. Nothing should be left to memory in this particular. Nor should caterpillars which are only presumably of the same species be placed in the same cage, as there are many allied kinds which are almost indistinguishable at sight, and a lack of exactitude here will vitiate one's observations.

Any one pursuing vigorously such a course of study and collection of native butterflies will be enchanted to see how fascinating the study is, how rapidly his collection grows, what an endless source of interest attaches to these humble but exquisite creatures, and into how many lines of real investigation his steps are tending. No one can undertake it without being himself the gainer by it, and without infusing others with his own ever-fresh enthusiasm.

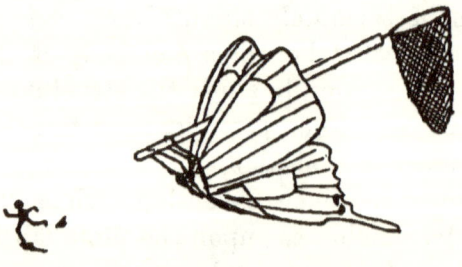

INDEX OF NAMES.

acadica, Thecla, 122
Achalarus lycidas, 166
Aglais, 36, 47, 54, 89
Aglais milberti, 89
Agraulis vanillae, 66
ajax, Iphiclides, 146
alcestis, Argynnis, 78
alope, Cercyonis, 110
Amblyscirtes samoset, 174
    vialis, 174
Anaea 37, 45, 55, 104
Anaea andria, 104
Ancyloxipha, 43, 52, 166
Ancyloxipha numitor, 166
andria, Anaea, 104
Angle Wings, 36, 47, 54, 82
Anosia, 34, 45, 55, 63
Anosia plexippus, 60, 63
Anthocharis, 40, 50, 58, 140
Anthocharis genutia, 140
Anthomaster, 44, 170
Anthomaster leonardus, 170
antiopa, Euvanessa, 90
Apatura celtis, 106
    clyton, 105
    herse, 105
    lycaon, 106
    proserpina, 105
aphrodite, Argynnis, 77
Araschnia prorsa, 16
archippus, Basilarchia, 102
Argus comyntas, 123
    eurydice, 108
Argynnis, 35, 46, 54, 76
Argynnis alcestis, 78
    aphrodite, 77
    atlantis, 76
    bellona, 72
    columbina, 81

Argynnis cybele, 79
    idalia, 80
    myrina, 74
arthemis, Basilarchia, 98
astyanax, Basilarchia, 101
atalanta, Vanessa, 87
Atalopedes huron, 174
atlantis, Argynnis, 76
Atlides halesus, 123
Atrytone, 44, 167
Atrytone logan, 169
    zabulon, 167
augustus, Incisalia, 116
Basilarchia, 37, 47, 55, 98
Basilarchia archippus, 102
    arthemis, 98
    astyanax, 101
    proserpina, 100
bellona, Brenthis, 72
Blues, 38, 48, 56, 123
Brenthis, 35, 46, 54, 72
Brenthis bellona, 72
    chariclea, 75
    freija, 75
    montinus, 75
    myrina, 74
brizo, Thanaos, 162
Brush-footed Butterflies, 25, 34, 45, 53, 63
caesonia, Zerene, 133
calanus, Thecla, 120
Calephelis borealis, 113
Callicista columella, 123
Callidryas, 40, 49, 57, 132
Callidryas eubule, 132
    philea, 133
    sennae, 133
Calpodes ethlius, 174
Calycopis cecrops, 123

INDEX OF NAMES.

cardui. Vanessa, 84
catullus, Pholisora, 164
celtis, Chlorippe, 106
Cercyonis, 37, 48, 56, 110
Cercyonis alope, 110
    nephele, 111
    pegala, 112
Charidryas, 35, 46, 53, 69
Charidryas ismeria, 70
    nycteis, 69
Chlorippe, 37, 47, 55, 105
Chlorippe celtis, 106
    clyton, 105
Chrysophanus, 39, 49, 57, 127
Chrysophanus americanus, 128
    epixanthe, 128
    hyllus, 127
    hypophlaeas, 128
    tarquinius, 130
    thoe, 127
Cinclidia, 35, 46, 53, 68
Cinclidia harrisii, 68
Cissia, 37, 47, 55, 107
Cissia eurytus, 107
    sosybius, 108
claudia, Euptoieta, 81
clyton, Chlorippe, 105
coenia, Junonia, 82
Coenonympha inornata, 112
Colias amphidusa, 135
    caesonia, 133
    chrysotheme, 135
    eurytheme, 132
    keewaydin, 135
    philodice, 134
comma, Polygonia, 95
comyntas, Everes, 123
Coppers, 39, 57, 127
Crescent Spots, 34, 45, 53, 66
cresphontes, Heraclides, 151
Cupido pseudargiolus, 125
Cyaniris, 38, 48, 56, 125
Cyaniris pseudargiolus, 18, 125
cybele, Argynnis, 79
Cynthia atalanta, 87
    cardui, 84
    huntera, 85
damon, Mitura, 118
Danaids, 34, 45, 55, 63
Danais archippus, 63
    erippus, 63
Debis portlandia, 109

Doxocopa herse, 105
    lycaon, 106
edwardsii, Thecla, 121
Emperors, 37, 55, 104
Enodia, 37, 48, 55, 109
Enodia portlandia, 109
Epargyreus, 43, 51, 59, 155
Epargyreus tityrus, 155
Epidemia, 39, 49, 57, 128
Epidemia dorcas, 128
    epixanthe, 128
    helloides, 128
epixanthe, Epidemia, 128
Erebia nephele, 111
Erora laeta, 123
Erycinids, 113
Erynnis, 44, 169
Erynnis attalus, 169
    manitoba, 169
    metea, 169
    sassacus, 169
    uncas, 169
eubule, Callidryas, 132
Eucheira socialis, 11
Eudamus proteus, 166
    pylades, 156
    tityrus, 155
Eugonia, 36, 47, 55, 92
Eugonia j-album, 92
Euphoeades, 42, 51, 58, 150
Euphoeades palamedes, 151
    troilus, 150
Euphydryas, 35, 46, 53, 66
Euphydryas phaeton, 66
Euphyes metacomet, 174
    verna, 174
Eupsyche m-album, 123
Euptoieta, 36, 46, 54, 81
Euptoieta claudia, 81
Euptychia eurytus, 107
Eurema, 40, 49, 57, 138
Eurema lisa, 138
    nicippe, 137
eurydice, Satyrodes, 108
Eurymus, 40, 50, 58, 134
Eurymus eurytheme, 19, 135
    interior, 136
    philodice, 134
eurytheme, Eurymus, 135
eurytus, Cissia, 107
Euvanessa, 36, 47, 54, 90
Euvanessa antiopa, 90

INDEX OF NAMES.

Everes, 38, 48, 56, 123
Everes comyntas, 17, 123
faunus, Polygonia, 94
Feniseca, 39, 49, 57, 130
Feniseca tarquinius, 130
Fritillaries, 35, 46, 54, 72
Gaeides dione, 131
genutia, Anthocharis, 140
glaucus, Jasoniades, 148
Goniloba tityrus, 155
Gossamer-winged Butterflies, 25, 37, 48, 56, 113
Grapta c-argenteum, 93
    comma, 95
    dryas, 95
    fabricii, 97
    faunus, 94
    interrogationis, 97
    j-album, 92
    progne, 93
    umbrosa, 97
Hair Streaks, 38, 48, 56, 113
harrisii, Cinclidia, 68
Heliconians, 66
Heodes, 39, 49, 57, 128
Heodes hypophlaeas, 128
Heraclides, 42, 51, 58, 151
Heraclides cresphontes, 151
Hesperia, 43, 52, 59, 165
Hesperia ahaton, 173
    centaureae, 166
    hobomok, 167
    montivaga, 165
    mystic, 171
    pocahontas, 167
    sassacus, 169
    tessellata, 165
    wamsutta 170
Hesperidae, 25
Heteropterus marginatus, 166
Hipparchia alope, 110
    andromacha, 109
    boisduvalii, 108
    eurytris, 107
    nephele, 111
huntera, Vanessa, 85
Hylephila phylaeus, 174
Hypatus bachmanii, 112
hypophlaeas, Heodes, 128
icelus, Thanaos, 163
idalia, Speyeria, 80
Incisalia, 38, 56, 114

Incisalia augustus, 116
    irus, 115
    niphon, 114
interrogationis, Polygonia, 97
iole, Nathalis, 139
Iphiclides, 41, 51, 58, 146
Iphiclides ajax, 17, 146
irus, Incisalia, 115
j-album, Eugonia, 92
Jasoniades, 42, 51, 58, 148
Jasoniades glaucus, 17, 148
    turnus, 148
Junonia, 36, 47, 54, 82
Junonia coenia, 82
    lavinia, 82
juvenalis, Thanaos, 161
Kallima, 24
Laertias, 41, 50, 58, 145
Laertias philenor, 145
Larger Skippers, 42, 51, 59, 155
leonardus, Authomaster, 170
Lerema accius, 174
    hianna, 174
Lerodea fusca, 174
Libytheinae, 26
Limenitis archippus, 102
    arthemis, 98
    astyanax, 101
    disippus, 102
    misippus, 102
    ursula, 101
Limochores, 44, 173
Limochores bimacula, 174
    manataaqua, 174
    palatka, 174
    pontiac, 174
    taumas, 173
liparops, Thecla, 119
lisa, Eurema, 138
Long Beaks, 112
lucilius, Thanaos, 158
Lycaena comyntas, 123
    epixanthe, 128
    neglecta, 125
    pseudargiolus, 125
    violacea, 125
Lycaenidae, 25
Meadow Browns, 37, 47, 55, 107
Meganostoma caesonia, 133
Megisto eurytus, 107
melinus, Uranotes, 117
Melitaea harrisii, 68

Melitaea marcia, 71
  nycteis, 69
  phaeton, 66
  pharos, 71
  tharos, 71
milberti, Aglais, 89
Minois alope, 110
  nephele, 111
Mitura, 38, 56, 118
Mitura damon, 118
montivaga, Hesperia, 165
myrina, Brenthis, 74
mystic, Thymelicus, 171
Nathalis, 40, 50, 58, 139
Nathalis iole, 139
  irene, 139
Neonympha canthus, 108
  cornelius, 112
  eurytris, 107
  mitchellii, 112,
  phocion, 112
nephele, Cercyonis, 111
nicippe, Xanthidia, 137
niphon, Incisalia, 114
Nisoniades brizo, 162
  catullus, 164
  ennius, 161
  icelus, 163
  juvenalis, 161
  lucilius, 158
  persius, 159
Nomiades couperi, 127
  lygdamus, 127
numitor, Ancyloxipha, 166
nycteis, Charidryas, 69
Nymphalidae, 25
Nymphalis arthemis, 98
  dryas, 95
  ephestion, 101
  faunus, 94
  j-album, 92
  lamina, 98
  milberti, 89
  ursula, 101
Nymphs, 34, 45, 53, 66
Oarisma poweshiek, 174
Oeneis calais, 112
  jutta, 112
  macounii, 112
  semidea, 112
oleracea, Pieris, 143
Oligoria maculata, 174

Orange Tips, 40, 50, 58, 140
Pamphila aaroni, 169
  cernes, 173
  leonardus, 170
  mandan, 174
  mystic, 171
  peckius, 170
  sassacus, 169
  zabulon, 167
Paphia glycerium, 104
  troglodyta, 104
Papilio, 42, 51, 59, 153
Papilio ajax, 146
  asterias, 153
  brevicauda, 154
  cresphontes, 151
  glaucus, 148
  marcellus, 146
  philenor, 145
  polyxenes, 153
  telamonides, 146
  thoas, 151
  troilus, 150
  turnus, 148
Papilionidae, 25
Pararge canthus, 108
peckius, Polites, 170
persius, Thanaos, 159
phaeton, Euphydryas, 66
philenor, Laertias, 145
philodice, Eurymus, 134
Phoebis agarithe, 140
Pholisora, 43, 52, 59, 164
Pholisora catullus, 164
  hayhurstii, 165
Phycanassa viator, 174
Phyciodes, 35, 46, 53, 71
Phyciodes batesii, 72
  gorgone, 72
  harrisii, 68
  nycteis, 69
  tharos, 17, 71
Pierids, 39, 49, 57, 132
Pieris, 41, 50, 58, 143
Pieris cruciferarum, 143
  frigida, 143
  napi, 143
  occidentalis, 141
  oleracea, 143
  protodice, 141
  rapae, 144
  vernalis, 141

INDEX OF NAMES.   205

plexippus, Anosia, 63
Pomues massasoit, 174
Polites, 44, 170
Polites peckius, 170
Polygonia, 36, 47, 55, 93
Polygonia comma, 95
   faunus, 94
   gracilis, 98
   interrogationis, 16, 97
   progne, 93
   satyrus, 98
Polyommatus comyntas, 123
   crataegi, 130
   epixanthe, 128
   lucia, 125
   porsenna, 130
   tarquinius, 130
   thoe, 127
polyxenes, Papilio, 153
Pontia, 41, 50, 58, 141
Pontia casta, 143
   oleracea, 143
   protodice, 141
portlandia, Enodia, 109
Potanthus omaha, 174
Prenes ocola, 174
   panoquin, 174
progne, Polygonia, 93
protodice, Pontia, 141
pseudargiolus, Cyaniris, 125
pylades, Thorybes, 156
Pyrameis atalanta, 87
   cardui, 84
   huntera, 85
   terpsichore, 85
   virginiensis, 85
Pyrgus montivagus, 165
Pyrisitia mexicana, 140
rapae, Pieris, 144
Red Horns, 39, 49, 57, 132
Rhabdoides cellus, 166
Rusticus scudderii, 127
   striatus, 127
sassacus, Erynnis, 169
Satyrodes, 37, 48, 55, 108
Satyrodes eurydice, 108
Satyrs, 37, 47, 55, 107
Satyrus alope, 110
   nephele, 111
   portlandia, 109
Semnopsyche diana, 82
Skippers, 25, 42, 51, 59, 155

Smaller Skippers, 43, 52, 59, 166
Sovereigns, 37, 47, 55, 98
Speyeria, 35, 46, 54, 80
Speyeria idalia, 80
Strymon, 38, 56, 113
Strymon melinus, 117
   titus, 113
Swallow Tails, 41, 50, 58, 145
Synchloe olympia, 141
Syrichtus communis, 165
tarquinius, Feniseca, 130
taumas, Limochores, 173
Terias lisa, 138
   nicippe, 137
Thanaos, 43, 52, 59, 158
Thanaos ausonius, 164
   brizo, 162
   horatius, 164
   icelus, 163
   juvenalis, 161
   lucilius, 158
   martialis, 164
   persius, 159
   terentius, 164
tharos, Phyciodes, 71
Thecla, 38, 56, 119
Thecla acadica, 122
   arsace, 115
   auburniana, 118
   augustus, 116
   borus, 122
   calanus, 120
   californica, 122
   costalis, 118
   cygnus, 122
   damon, 118
   edwardsii, 121
   falacer, 120
   favonius, 117
   henrici, 115
   humuli, 117
   hyperici, 117
   inorata, 120
   irus, 115
   liparops, 119
   lorata, 123
   melinus, 117
   mopsus, 113
   niphon, 114
   ontario, 123
   smilacis, 118
   souhegan, 122

INDEX OF NAMES.

Thecla strigosa, 119
   titus, 113
thoe, Chrysophanus, 127
Thorybes, 43, 51, 59, 156
Thorybes bathyllus, 158
   electra, 158
   pylades, 156
Thymele tityrus, 155
Thymelicus, 44, 171
Thymelicus aetna, 172
   brettus, 172
   mystic, 171
   numitor, 166
titus, Strymon, 113
tityrus, Epargyreus, 155
troilus, Euphoeades, 150
Typical Butterflies, 25, 39, 49, 132
Uranotes, 38, 56, 117
Uranotes melinus, 117
Vanessa, 36, 47, 54, 84
Vanessa antiopa, 90

Vanessa atalanta, 87
   c-album, 95
   cardui, 84
   coenia, 82
   comma, 95
   faunus, 94
   furcillata, 89
   huntera, 10, 85
   interrogationis, 97
   j-album, 92
   milberti, 89
   progne, 93
Whites, 41, 50, 58, 141
Xanthidia, 40, 49, 57, 137
Xanthidia lisa, 138
   nicippe, 137
Yellows, 39, 49, 57, 132
zabulon, Atrytone, 167
Zerene, 40, 50, 58, 133
Zerene anthyale, 134
   caesonia, 133

www.ingramcontent.com/pod-product-compliance
Lightning Source LLC
Chambersburg PA
CBHW031828230426
43669CB00009B/1260